BECAUSE THE CAT *Purrs*

Also by Janet Lembke

From Grass to Gardens

The Quality of Life

Skinny Dipping

River Time

BECAUSE THE CAT *Purrs*

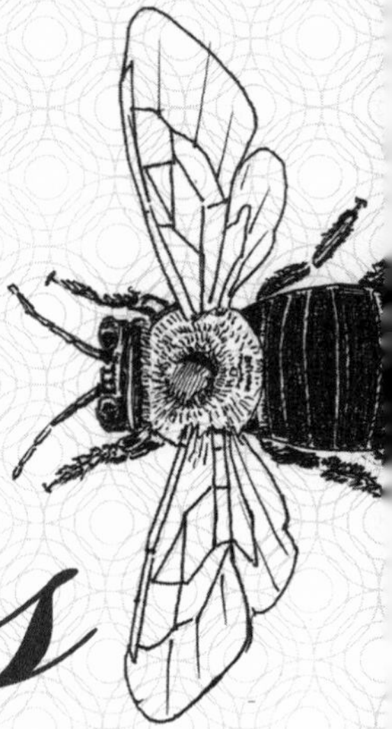

How We Relate to Other Species
and Why It Matters

janet lembke

Skyhorse Publishing

for my great-grandson
Ethan Isaac Neblock

Skyhorse Publishing books may be purchased in bulk at special discounts for sales promotion, corporate gifts, fund raising, or educational purposes. Special editions can also be created to specifications. For details, contact Special Sales Department, Skyhorse Publishing, 555 Eighth Avenue, Suite 903, New York, NY 10018 or info@skyhorsepublishing.com.

www.skyhorsepublishing.com

10 9 8 7 6 5 4 3 2 1

Library of Congress Cataloging-in-Publication Data

Lembke, Janet.
Because the cat purrs : how we relate to other species and why it matters / Janet Lembke.
p. cm.
ISBN 978-1-60239-235-9 (hardcover : alk. paper)
1. Human-animal relationships. 2. Animal behavior. I. Title.

QL85.L36 2007
508—dc22

2007042144

Printed in the United States of America

The real,
which is perfectly simple,
and supremely beautiful,
too often escapes us,
giving way before the imaginary,
which is less troublesome to acquire.

—JEAN-HENRI FABRE

Contents

Acknowledgments

Heartfelt thanks go to all of the people who are named in these stories. Others, who do not appear in the text, deserve special mention:

Jeffery Beam and David Romito for replying in a trice to my inquiries about the meanings of scientific names.

Eric Day and Richard Fell of the Cooperative Extension Agency at Virginia Tech for helping out, respectively, with identifications and apiculture.

Dr. Julie Glenn for demystifying *Peromyscus*, "Mouse in Boots."

Dr. Julian Ketley for his knowledge of *Campylobacter jejuni*.

Karen Cochran and Carroll Lisle, friends, for their stories.

Last but hardly least, my editor Nick Lyons warrants much gratitude. Without his nudging, I would have been gardening rather than writing a book.

Them and Us

{ *An Introduction* }

I begin, inevitably, with the cat. Those of us who are possessed by them know full well that they often illustrate cognition, the sorting through of the information that presents itself thus: Shall I leap upon the kitchen counter, or shall I not? Let me introduce my domestic shorthairs, both neutered tomcats. Omega is a brown mackerel tabby (mackerel means that his tiger stripes are continuous, not broken), while Alpha is black with a white bib and mustache and a small black goatee; both live strictly indoors (no need for fights with feral cats or bird-slaughter or exposure to fleas, ticks, and feline leukemia). Although they did not know each other before they came to live with me, they bonded immediately.

Now, years later, they wrestle, chase each other thunderously up and down the stairs, eat from the same dish at the same time, and groom each other. Their cognitive processes seem to be self-referential and deal mainly with what is advantageous or desired—food, exercise, attention, and the maintenance of routine. Like all cats, neither has a point of view. And, though self-aware mainly at those rare moments that a tail is stepped on, they do not recognize themselves in a mirror, nor can they behave like philosophers and conceive of thinking itself as an object for thought. But they certainly possess a social awareness that extends to their own kind, to the other animals that enter their territory, and to people. They know I am not one of them but consider me as an integral part of their group. Social awareness is not, however, the force that propels Alpha headfirst against the kitchen window in an attempt to catch one of the birds at the feeder. The impetus there is that he's been hardwired as a hunter. The Omega cat is epileptic; and during a seizure, Alpha will look on with obvious puzzlement: routine has gone most peculiarly askew. And if Omega must spend part of the day at the vet's, Alpha, usually soft-spoken, meowls loudly. As for being conscious, not just cognitive, it's fair to assume that cats are as aware of what goes on around them as is a baby of six or so months. They are certainly capable of learning. Not only do they answer to their names but they have also figured out how to use body language or vocalization to tell me what they want. And Omega is a fetch cat; he'll bring me a straw, which I toss out, whereupon he retrieves it, and we play the game again for another four or five innings. In general, cats are either social or asocial, pro or con, not neutral (though they have been known to sit on the fence). But I confess to bias and confess, as well, that having never been a cat, I cannot know what it is to be one. In the course of these stories, you'll meet a cat named Sophie, whose behavior has made me think about the odd bargain that we make with *Felis sylvestris catus*, the domestic cat.

I do know that the relationship of person and pet is hardly neutral. It is commensal, a term that means "together at the table." This kind of symbiosis has been characterized as a Plus-Zero arrangement, in which one party, usually the smaller, benefits, while the other is not affected. But with pets and people, that is only a partial truth. The former certainly enjoy material advantages—food, water, treats, toys, shelter, exercise, and doctoring, while the latter reap intangible rewards, like companionship and sheer pleasure. Nonetheless, though both parties profit from their association, though it seems to be a Plus-Plus situation, it cannot be characterized as mutualism. In that type of symbiosis, the very survival of each party depends upon the other. Pollinator and plant, honeybee and almond tree, hummingbird and scarlet runner bean—one needs nectar and/or pollen for itself and for its young, while the other needs the brush of pollen-laden legs or bill against its stigma so that it can set seed.

Most of the time, we human beings go about our lives without paying one whit of attention to the modes in which we intersect with others. There's no need to bother about *Them*, for they don't interfere with *Us*. Sometimes, our intersections with Them produce win-win situations. My next-door neighbor and the squirrel that learned to eat peanuts out of his hand upped their commensal relationship by several notches. My own connection to the man-squirrel duo took the form of finding peanuts sprouting in my raised-bed gardens the following year; my only regret there is that I yanked the plants before I realized that they were peanuts; the tell-tale shell had been well buried. As for insects, my unasked-for association with carpenter bees has turned out to be decidedly win-win; more about that later. But sometimes, without the slightest intent, we do interfere with other lives and so provoke retaliatory strikes. Run a lawnmower over a hole in the ground, and a swarm of yellow jackets rises, bayonets at the ready. All unaware, approach a blue jay's nest, and the bird swoops down to stab its not inconsiderable beak into an innocent head. Pat a strange dog, suffer a bite. Most of the time, however, the equation works the other way around. It's not that we interfere with them, but rather that they willy-nilly affect us, not always for the best. Sometimes, the results, though noticeable, are gentle: a vine with small red flowers pops up in the garden; it's a wildling but it's beautiful—let it stay. But sometimes, jarring collisions occur between Them and Us, and they may happen, bolts from nowhere, when we are simply pursuing our ordinary, quotidian tasks and entertainments; we may, in all innocence, be gardening or driving to work, eating out at a restaurant or bird-watching.

In such cases, species once neutral or commensal do not back off but follow the mandatory paths mapped out by their instincts. Find food. Defend territory. Seek a mate. There will be weevils in the flour, feral cats staking out the bird feeder, voles demolishing daffodil bulbs, and bacteria that roil our guts and turn us into wretches. When we do become losers, neutrality and commensalism fly out the window, and other symbioses come right quickly to the fore. The confrontations become Them *or* Us.

One of our response is predation, a Plus-Minus interaction, in which we actively pursue the foe with the intent of giving it a *coup de grâce* or, at very least, disabling it. We set traps; we throw rocks; we shoot with bullets or BBs. I know a man who kicks groundhogs; you'll meet him later. Another kind of symbiosis that may take command of our responses to Them is amensalism, "away from the table." We shove Them as far away from sharing our goods as we possibly can; we aim for obliteration. Amensalism appears in two Zero-Minus guises, in which one party loses, while the other remains neutral, gaining no profit except for getting rid of the foe. They are competition and antibiosis, both of which can be as lethal as predation. For me, the first form means wielding a hoe or three-pronged hand fork: damned if I'll allow those common mallows, wild violets, and dandelions to compete for space amid my vegetables, herbs, and daylilies. Their competition triggers mine. Antibiosis, amensalism's second form, means "against life," and it posits the use of poisons meant to kill—rodenticides against mice and rats; herbicides against weeds; rotenone and Bt against cabbage worms; flames against the tent caterpillars, penicillin and other drugs against all manner of bacteria and viruses that make us sick. Plants and fungi also engage in antibiotic warfare so that they can keep their territories and the nutrients therein all to themselves. Black walnuts exude juglone, a toxin that only a few sturdy plants can withstand; tomatoes, peppers, and potatoes are not among them. My horseradish plants manufacture an exudate that increases the alkalinity of the soil, a maneuver that not only repels competitors but also acts to promote the plants' own growth.

One other form of Them and Us symbiosis is parasitism—the tick and the mosquito sucking up our blood; *Plasmodium,* the malarial protozoan, attacking our red blood cells; the fungi, like ringworm and athlete's foot, that infect our skin. Here, Plus always represents Them, while Minus invariably characterizes Us (unless, of course, vampires like Dracula and Lestat are brought into the equation, in which case, it is Us and Us). It may be comforting for Us to know that in the case of ticks, at least two forms of microscopic parasites infest Them.

*

Them and Us—the operative word is *and.* We share this earth with an uncountable multitude of other lives, each belonging to one of the three domains and six kingdoms into which scientists have currently (and disputatiously) sorted living things. Much of the time our relationships—person with mammal, bird, bug, vine, and even bacterium—are neutral, neither impinging on the other's way of being and moving about in the world. Instinct guides all of us, some of us to the exclusion of conscious thought and decision-making. I am convinced, however, that not just instinct, not just cognition, which is a neural processing of information, but also intelligence, thought, and the ability to make choices are hardly restricted to humankind and our anthropoid kin but extend to other animals, like dogs, cats, parrots, elephants, and bees, and maybe even to groundhogs. I have observed a dog and a great blue heron at play, teasing each other, and crows jeering, making sport of someone who would shoot them.

Many researchers have focused on animal intelligence, and their findings amply support the presence of awareness, communication, and decision-making in many creatures. An abundance of animals—dolphins, cormorants, Dalmatian pelicans, eagles, South American giant otters, hyenas, and lions, to mention only a few—engage in cooperative hunting. Beavers and birds are capable of elaborate engineering when it comes to dens and nests. Not only monkeys and apes use tools; some ants fashion sponges to carry semi-liquid food back to their nests, certain assassin bugs make tools to capture termites, and Darwin's finches pry insect larvae and pupae from dead branches with twigs or cactus spines. Bees dance to convey information on the location of nectar and pollen sources, and choosing between a multitude of tasks, they can start one—tending the queen, for example—and change their minds to embark on, say, foraging or feeding the larvae. In their colonial situation of hundreds to thousands of insects living wing by wing, it is imperative that the hive's inhabitants respond sensibly to the colony's changing needs and that all the tasks that need doing are, indeed, done. The survival of both individuals and the group depends on cooperation, but within that *sine qua non* requirement, each member of the group is free to choose among the many available jobs. No one is unemployed, and no particular bee is the boss.

Birds, too, engage in complex behavior that points to careful consideration of a task or a problem. One study reports that a crow fashioned a tool by bending wire into a hook for grabbing food. And the "cognitive potential" of birds, as James R. Gould and Carol Grant Gould have phrased it, allows many of them to build architecturally complex nests, which in turn permit them to raise their young more successfully in an expanded niche. (The Goulds are, respectively, an ecologist-cum-evolutionary biologist and a science writer.) Not just cognition but the exercise of choice in birds is certainly evident in the structures built by male bowerbirds. They are not nests but rather elaborately decorated stages built for courting and mating, and the female birds are selective to the point of being persnickety: this bower lacks pizzazz; and so, most likely, does the bird who built it; so, I'll keep on looking. The Goulds characterize the bowers and their architects this way:

> At the cognitive level, these creations are the most complex seen in birds. Only beavers and humans undertake work with more steps and greater flexibility in design, materials, and execution. Bowers require a juggling of multiple perspectives and parameters. And because males are competing desperately to reproduce, with only this artifact to save them from genetic oblivion, selection has strongly rewarded the sorts of mental tools we generally associate with intelligence.

It might be said that constructing a bower is a genetically programmed act, but the many choices that the bird must make—the arrangement of building materials, the selection of decorative feathers and flowers, the defense of his bower from thievery while he attempts to steal from others—all point to an active exercise of intelligence. So much for being bird-brained.

What about our closest primate relatives, the chimpanzees and bonobos? When it comes to them, there is no question about their ability to think and to communicate with grunts and hoots, grimaces, smiles, and gestures. A begging hand says, "Give me food," or, "Give back the food that you stole." A young bonobo's upraised arms, one hand bending back the fingers of the other, say, "I want sex." And individual members of troops have invented meaningful gestures that the others learn and use.

*

The domains and kingdoms to which ticks and their parasites, ringworm, horseradishes, cats both feral and domestic, and all the rest of Them and Us have been assigned are worth a brief look. All of us are here, willy-nilly, aboard planet Earth. (Mind you, we are the only species that makes judgments for or against other species; we are the only ones who insist on sorting life-forms into discrete categories.)

The domains are three: Bacteria, the rod-like life-forms; Archaea, the ancient life-forms; and Eukarya, life-forms with nucleus-containing cells. Inhabitants of the first two domains are prokaryotes—creatures "before a nucleus"—and they are misleadingly named, for Archaea are not more ancient than Bacteria. Both developed from the aboriginal forms of life, which arose in the electrical storms that zapped the primordial seas and triggered chemical reactions that fused some of the elements into creatures that could reproduce. (I have long had a vision of those storms—sperm-like lightning inseminating the world-egg.) All the inhabitants of both domains are single-celled circles of DNA. Though they lack a nucleus, they contain specialized organelles which exist within each cell to receive stimuli and respond to them. Among these "little organs" are light-sensitive spots, sensory hairs, and filaments that contract like muscles to move the cell through its fluid medium.

The Bacteria belong to the Kingdom Eubacteria, a designation that some have casually translated as "true bacteria," but the Greek syllable *eu-* actually means "good" or "abundant." And bacteria are indeed the most abundant creatures on this earth. A single bacterium can generate ten million more between one midnight and the next. They come in two varieties, aerobic bacteria, which use oxygen for respiration, and anaerobic bacteria, which live in an oxygen-free environment. Both varieties have their share of organisms with which we can live in commensal comfort and of those that lay us low in Plus-Minus symbioses that favor Them over Us. We can thank anaerobic bacteria for the fermentation that produces cheese, sauerkraut, and wine. We can also thank them for the toxin used in Botox but damn them for the botulism caused by the same toxin. (It's worth mentioning that not all Botox is used for facultative purposes like smoothing out the wrinkles in one's face. It greatly relaxes facial tics and the spasms that some people suffer to the point of not being able to open eyelids that were involuntarily clenched shut.) Other bacteria fix nitrogen in the soil, synthesize vitamins in mammalian guts, and digest the petroleum cast upon the waters by oil spills. One bacterium, *Mycobacterium vaccae*, "fungus-bacterium of the cow," which lives in the soil, warrants special mention: researchers have discovered that it triggers brain cells to release serotonin, a chemical with many functions, one of which is to act as a natural antidepressant regulating moods. So, gardening—touching earth—is good for us. We also damn bacteria for a host of diseases, from strep throats and digestive upsets to cholera, tuberculosis, syphilis, and bubonic plague. A few of the multitudinous Eubacteria will make an appearance later in this book.

The single-celled ancient ones belong to a kingdom with the same name as their domain: Archaea. Their DNA, the structure of their cell walls, and the composition of their plasma membranes differentiate them from the Eubacteria. The plasma membranes serve as filters that let in desirable substances while keeping all the others out. The ancients come in four varieties, two of them adapted to extreme conditions. The thermophiles, "heat lovers," which thrive at temperatures up to 230 degrees Fahrenheit, snuggle into habitats like hot springs, geysers, and oil wells. The halophiles, "salt lovers," make their homes in places that are far saltier than the oceans; they may be found in Utah's Great Salt Lake and the Dead Sea, where the waters take on a reddish tinge from the pigments in these microbes. They also make themselves at home in very salty foods, like sausages, salt pork, and salt fish. The third group is that of the methanogens, the methane-makers, which live in anaerobic environments and produce the gas for which they are named. The fourth is that of the ubiquitous mesophiles, the "middle-lovers," which, like bacteria, flourish at moderate temperatures. People are inhabited by one variety of the ancient ones—to be specific, the methane-makers. We carry them in the wet places of our bodies—alimentary tract, mouth, vagina. Nor are we the only creatures who house them; they may be found within many other animals, including the hindgut of the cockroach. Though we furnish methanogens with homes, the relationship between Them and Us seems commensal. No evidence yet exists that implicates any of them as causes of disease. (It may be that we've been blaming the wrong prokaryotes for our ills.) Nonetheless, they are definitely responsible for methane production in our guts and for rip-roaring sales of products like Beano.

We arrive at the domain of the Eukarya, life-forms abundantly blessed with nuclei, though the domain does contain some one-celled organisms. But even those single cells are larger and more complex than those of the prokaryotes. Like the prokaryotes, we have organelles, two of which are mitochondria and chloroplasts, the latter found only in plants and protoctistans. Educated theory has it that chloroplasts were once free-living "green shapes," probably descended from blue-green algae, while mitochondria, "thread granules," most likely originated as Proteobacteria—"first bacteria," a group that includes the infamous *Salmonella* and *E. coli*. These organelles are called endosymbionts—symbionts that live commensally inside a larger unit. All of us eukaryotes have nuclei with our DNA arranged in chromosomes.

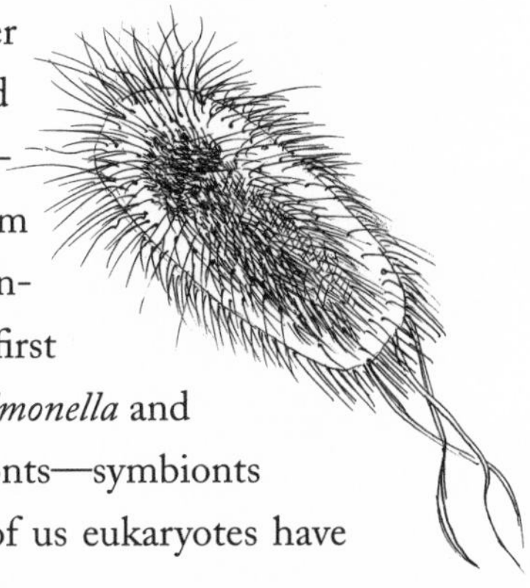

The domain of the Eukarya comprises four kingdoms: Protoctista, Plantae, Fungi, and Animalia. The Protoctista, the "first established," constitute a culch pile, in which creatures that can't be fitted neatly into any other eukaryote kingdom are heaped up willy-nilly. As the name indicates, they are the earliest of the Eukarya. (Some scientists call this kingdom Protista, the "first ones.") Primarily—but not invariably—single-celled, they include amoebae; paramecia; simple red, green, and brown algae, including kelp; and slime molds. The single-celled microbe that causes malaria is a protoctistan, and dinoflagellates, microorganisms that propel themselves through water with whirling, thread-like filaments, also belong to the same kingdom. Some of these minuscule dinoflagellates are predatory hunters able to secrete neurotoxins that kill fish, on which they then proceed to feast. And it is they that cause red tides, containing toxins that can affect us severely.

On the other hand, the slime molds, despite their oozy name, are benign in respect to their dealings with Us. The word "mold" in the slime mold's name might lead you to think that they should be classified as denizens of the kingdom Fungi. Once upon a time, they were, until it was discovered that they, unlike the fungi, have no cell walls when they are clumped together in a vegetative state—that is, when they arrive at their moment of quiescence before active spore production begins. Each slime mold starts as a single spore, which grows into a cell and then seeks others of its kind. Finding one another, they cohere and form a multinucleated glob of protoplasm, a colonial arrangement, which has the power to move. Though some slime molds are plain white, others wear fine colors, like bubble-gum pink, brick red, lemon yellow, and cobalt blue. Some of these aggregations, depending on the species, have smooth exteriors, while some look covered with small bumps. A somewhat wrinkled-looking yellow slime mold the size and shape of half of a banana appeared not long ago in my garden at the base of a front-yard daylily. When the bacterial food in one location has been exhausted, the plasmodium moves onward almost imperceptibly, covering a fraction of a centimeter in an hour. The goal is to find a suitable place to produce spores. Likely, the slime mold beside the daylily had reached this fruiting stage. Food and reproduction—the goals of a slime mold are those of any living creature.

Born in the ocean, most of the Protoctista have stayed in water, although the slime molds climbed ashore. But Plantae, the plants, were the first to make themselves thoroughly at home on dry land and to colonize every crevice in which they found it possible to grow. The ancestral plants were primitive green algae that had played a successful survival-of-the-fittest game, in which single-celled seagoing organisms were tossed ashore by waves and left to perish in the noonday sun. Those organisms were creatures of the water, which furnished them with food, and they released their eggs and sperm into this fluid medium where the two could come together to form minute egg-like zoospores—life-spores. But some of the beached spores were tenacious, able to withstand short periods of aridity and still live on. And the places upon which they were cast away were not like the terrestrial places of today; instead, composed of sand, mud, and clay, that ancient terra firma lacked organic matter. It was the plants themselves that made soil as we know it, for when they died, they left their bodies strewn on those unforgiving substances. Bacteria and fungi evolved that could break down organic plant matter to create soil that was friendly not only to roots but to living things like worms.

The metamorphosis of algae to plants began some 450 million years ago, as the Ordovician period rolled over into the Silurian. Out of necessity, the survivors came up with clever stratagems for staying alive on dry land. Their innovations include a cuticle, or waxy covering, to keep leaves from drying out; stomata, or pores, for taking in carbon dioxide and releasing oxygen; vascular structures for the transport of nutrients from the soil and from the air; and pollen for insemination. (Later, in these stories, a pollinator will appear.) First on the high-and-dry scene were the mosses and ferns. Plants underwent an explosive diversification in the Devonian period, which began 408 million years ago and lasted for 46 million years, when the Mississippian period began. During the Mississippian, the gymnosperms, the "naked seeds," made their first appearances—"naked" because the seeds are small and not encased in a fruit. Oh, the conifers! They are plants that brave fortune by leaving their seeds bare—venerable cycads with palm-like leaves and a stout trunk, Christmas trees from white pines to Norway spruce, and all the cone-bearers in between, including the redwoods, and ginkgos, the leaves of which are shaped like an open fan and the fleshy seeds of which emit a nose-wrinkling reek when they're inadvertently squished underfoot. The angiosperms, the "contained seeds" with protective coverings, literally came to flourish—to flower, that is—in the Early Cretaceous some 120 million years ago. The encounters between Them and Us are manifold. They supply us with food, from berries and nuts to apples and peaches; they provide lumber for our dwellings and furniture; giving shade, transpiring moisture, they cool our mutual environs; they absorb carbon dioxide, a notorious greenhouse gas. We should be on our guard against others, like poison ivy and poison sumac with their itch-inciting toxin, urushiol; like the upstart weeds that overrun our lawns and gardens; like ailanthus and melaleuca trees, exotic invasives that crowd out native plants. In these stories, you'll meet two members of the Plantae: morning glories and red maple.

Like the Plantae, the kingdom of the Fungi evolved on land. This kingdom comprises the Phycomycetes, Ascomycetes, Basidiomycetes, and Fungi Imperfecta. The names translate, respectively, as "weed fungi," "skin fungi," pedestal fungi," and "imperfect fungi." The large weed-fungi clan is responsible for the pin molds that sprout on food and leather, for bread molds, and for the Irish potato blight that began in 1845. On the positive side, some of them assist us in the manufacture of all forms of potable alcohol. The skins, which enclose their spores in skin-like sacs, bring us many boons, like baker's and brewer's yeasts, the edible cup fungi that grow on wood, and the delectable morel mushrooms. Pedestal designates not only the familiar toadstools and mushrooms that wear caps upon their stalks, but also the bracket fungi and puffballs. Some are succulent, while others, like the death angel, produce lethal toxins. The imperfects, so named because they are not capable of sexual reproduction, aid us by helping to ripen Stilton, Roquefort, Gorgonzola, and other cheeses. They also assail us by causing diseases like athlete's foot, ringworm, jock itch, and nail fungus. Some imperfects can affect human lungs; some can kill. Each group of fungi, from the weeds to the imperfects, contains three variants, distinguished one from the others by what they eat: parasites, obliged to eat living tissue; saprophytes, eating only dead matter; and opportunists, which can go either way, depending what is in the natural larder. The inborn goal of all fungi, no matter what their feeding habits, is to clean up the world by decomposing organic matter, be it alive or dead, and absorbing it. Otherwise, the planet would long ago have been smothered by its own wastes.

People and the other multitudinous members of the kingdom Animalia are close to the Fungi in evolutionary terms. That may seem a most peculiar and unlikely state of affairs, but the existence of a phylogenetic relationship is given substance in a number of ways. For one, neither fungi nor animals have chloroplasts. We do share morphological and structural cellular features, and our biochemical pathways are similar. (Ah, my beloved though distant cousins, the morels and shiitakes! Oh, my spurned relatives, the refrigerator molds and fungal itches!) But while the fungi evolved on land, only one type of animal did so—the insects. The ancestors of the animals who are now terrestrial creatures—birds, reptiles, worms, mammals, and the motley others—crawled out of the sea. But some, from crustaceans and corals to fish, dolphins, and whales, never abandoned the fluent waters. Still others—turtles and snails, for example—have divided loyalties, with some genera living in water and some ashore. And what a higgledy-piggledy creeping, crawling, squirming, swimming, strutting, soaring gallimaufry we are! On occasion, we not only encounter one another but also become right royally entangled. Chance, often as not, transforms neutrality into a type of symbiosis, sometimes respectful, sometimes not. Nor is the entanglement always physical. For me, it often consists of a stiff poke at my curiosity bone. Only investigation mitigates the punch. So, in the realm of the Animalia, I here explore the ways in which we may find our lives taken over for better or worse, forever or for a little while, not by exotica but by such perfectly ordinary animals as cats and mice, white-tailed deer, groundhogs, turtles, snails, and house sparrows.

*

As always on my journeys into the nature of things, I have found good (if ghostly) companions, whose curiosity and wonder not only led them to conduct keen inspections of their environs but also to write copious records of the things and phenomena that they observed. You'll meet them more fully in the stories, but I'd like to make brief introductions here. From the centuries B.C. come the Greek historian Herodotus, the Roman farmer-statesman Varro, and the Roman poets Virgil and Catullus to give their observations in prose and verse. Pliny the Elder offers his observations from the first century A.D. English poet John Skelton enters from the late 1400s and early 1500s, while British botanist John Gerard arrives from the late 1500s. The 1700s are prodigious, sending me help from Swedish botanist Peter Kalm, English naturalist Mark Catesby, English gentleman surveyor John Lawson, Scottish poet Robert Burns, and President Thomas Jefferson. John James Audubon, Walt Whitman, and French entomologist Jean-Henri Fabre come on the scene from the 1800s; Nobel laureate Maurice Maeterlinck and American botanist Harriet Keeler materialize just slightly later, from the early 1900s. The information and insights that they hand on cover subjects that range from bees and morning glories to cats and mice. I marvel at their energies and rejoice in their company.

BECAUSE THE CAT *Purrs*

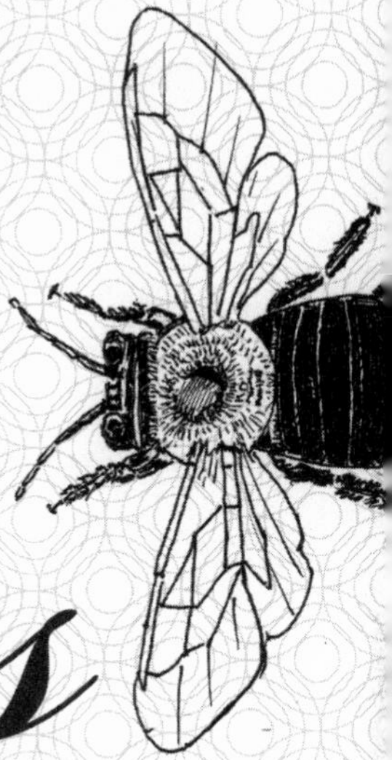

ONE

Sophie

{A Cautionary Tale}

Sophie looks like a cloud of sooty gray smoke. She's longhaired and spooky, with ice-cold eyes. And she's needy. A family living cater-cornered across the street adopted and named her several years ago, along with a skittish, coal-black cat that they dubbed Malcolm. My neighbors keep bowls of food and water on their front porch. They also went so far as to trap both cats and cart them to the vet for an examination and the appropriate shots. Malcolm was neutered, but, as it happened, Sophie had already been spayed. Once upon a time, she'd been somebody's darling. Occasionally, she goes inside the neighbors' house but soon thereafter returns to her chosen domain, the street, which she cases daily, strolling uphill and down again, sun, rain, or snow, until she feels a need to nap. Then, any porch will do, though she seems to prefer porches with cushioned chairs. One day, to my surprise, she introduced herself to me with a soft meow and a territorial rub against one of my porch pillars. I held out a hand, and—oh my!—claiming it, she allowed me to stroke her. Beneath the cloud of hair, she is little more than skin and bones. No meat on her, but from deep within, she produced a purr. Now, asking not for food but for attention, she makes occasional calls when I sit upon my porch. My own cats, indoor creatures both, stand at a window and watch with cocked heads, fully alert. I do not know what goes through their brains, but they make no moves, thank goodness, to dash outside and seem content to stay right where they are.

Sophie and Malcolm are not the only street cats that we see daily, though some are not strictly street cats. They might better be known as backyard cats. One large cream-colored tom often takes his repose beneath my Norway spruce. (How do I know that he's a tom, and intact? Because he's jowly and his cheeks bulge, as if he'd puffed them out.) The neighborhood also has at least one ex–street cat, Leroy, who prowled up and down in his kittenhood. As cats go, he was a sociable creature, readily allowing people to handle him. My young neighbors,

whom you will meet at greater length in the red-maple story, took him in just at the point that his testicles had descended. Next stop: the vet. He's now an indoor-outdoor cat. Although his people would like him to stay inside all the time, he gets peevish at being confined and strikes out at his housemate, Ivy, the dog. If Leroy is in my yard, he flees at the sight of me. Why? Not long ago, his female person had no luck when she tried to put him into a carrier for a trip to the cat-sitter. Frantic, she arrived on my doorstep to see if I'd care for Leroy while she and her husband vacationed far away. As she was showing me the location of his food and litter box, he came into range. I caught him and plopped him into the top-loading carrier. He's not about to let me catch him again. But Leroy's adoption is a success story for reasons that I'll come to shortly.

A clowder of unadoptable, irredeemably feral cats lives at my brother's mill, where he formulates and grinds feed-mixes—corn for energy, soy for proteins—to nourish all manner of farm animals, from cattle to llamas. The cats arrived not long after the mill opened. There's a rule at work here: where there is grain, there are rodents. To a cat, their presence means food and the thrill of the hunt. The current number of cats at the mill is anybody's guess, but it's safe to say at least a dozen. Nor need they depend solely on rodents for their suppers. My brother has set out bowls and keeps them filled with cat chow. Their relationship to my brother and the mill's employees is commensal. While the people are neutral, the cats depend on human activity for their sustenance, be it chow or the mice attracted by the feed. Nonetheless, all of them, no matter what their markings—orange tabby, black, broken white—are small and skinny. Anyone so foolish as to catch one finds that he holds a squirming, protesting creature with its claws unsheathed. These cats most successfully fend off any attempts to docile and domesticate them. It is as if they exercise volition or as if they have taken a most solemn vow: they shall not be anyone's pets.

✻

Not all domestic cats were meant to be pets. At least one, Bast by name, was—and maybe still is—a goddess. She entered the world during Egypt's second dynasty, some five millennia ago, appearing first as a woman with the head of a lioness. But she was gradually tamed and came to wear the head of a cat. Eventually, she was represented wholly as a domestic shorthair, from her large, perked ears to the tip of her tail. In any form, she was worshipped, sometimes extravagantly. The Greek historian Herodotus (fifth century B.C.) has described in detail how devotees came annually in spring to Bast's sacred city, Per-Bast, in the Nile delta. Men and women by the tens of thousands boarded barges, the women clattering castanets and some of the men playing flutes, and when they arrived, Herodotus says, "They celebrate the festival with elaborate sacrifices, and more wine is consumed than during all the rest of the year." And just what did Bast, sometimes called Bastet or Ubasti, signify to those who adored her? Her name means "devourer," and as such, she had responsibility for defending the pharaoh from enemies, for slaying them as a cat would slay and consume a mouse. She also showed her tender side as the protectress of pregnant women, caring for them as a queen would for her kittens. Herodotus also has much to say about house cats. When queens bear kittens, they avoid the toms, but the toms, "deprived of their satisfaction," have most ingeniously figured out how to regain what they have been denied—they steal the kittens, sometimes killing but not eating them, and the queens, deprived of their kittens, will again seek mates. As for their human commensals, Herodotus writes:

> What happens when a house catches fire is most extraordinary: nobody takes the least trouble to put it out, for it is only the cats that matter: everyone stands in a row, a little distance from his neighbors, trying to protect the cats, who nevertheless slip through the line, or jump over it, and hurl themselves into the flames.

He attests to the esteem in which cats are held by noting that the tenants of a house in which a cat has died a natural death express their grief by shaving off their eyebrows. (At a dog's death, the tenants shave their entire bodies, including the head.) And he tells us that dead cats were taken to Per-Bast (which the Greeks called Bubastis), where they were mummified and buried in "sacred receptacles." Likely, some of them were sacrificial victims. Archaeologists have since uncovered just such a cat cemetery in the ruins of Per-Bast, a site now called Tell-Basta. Nor was that the only Egyptian graveyard reserved for cats; many others have been found.

Once upon a time, some twelve millennia ago, all cats were wild. The evolution of the domestic cat has been demystified from studies made of the mummies that these graveyards have yielded. Noted science writer Sue Hubbell says, in her 2001 book *Shrinking the Cat*, "So many cats were mummified that toward the end of the nineteenth century boatloads of excavated mummies were exported from Egypt to England as ballast on the home voyage of commercial ships." (In the nineteenth century, human mummies were used as fuel for locomotives in the Middle East and were exported to England, where they served the same purpose in factories.) Ms. Hubbell points out that the Egyptians had engaged in genetic engineering thousands of years before anyone had conceived of genes. They'd tinkered with the North African version of the ancestral wildcat, a large, rangy, striped, shorthaired beast with gray or brown tabby markings, until it became a smaller and more docile animal for which the Egyptian name, then and now, is the onomatopoeic *mau.* Natural selection had nothing to do with the pussification of the wildcat. Rather, human beings bred the wild stock selectively for traits, from size and coat color to temperament, which would make them gentle, companionable, pleasing to our senses, and willing to live inside. The mummies show that as the size of the animal's body was reduced, its brain also became smaller. Imagination posits that its brain must have lost a bump of suspicion, along with a node directing it how to fend for itself. In actuality, the domesticated cat had lost some thirty percent of the neurons associated with vision, and its adrenal glands had also become smaller. In other words, puss, no longer a dedicated hunter, did not need acuity of vision to find food nor quickness of response to prey or danger. (It's noteworthy that a cat does not need a fully functioning brain in the first place; a cat with a brain injury that would turn a person into a vegetable continues to behave as usual—meowing, begging for food, using a litter box, and purring—because its autonomic nervous system kicks in and dictates normal behavior.)

And so, *Felis sylvestris*—the "cat of the woods," the "undomesticated cat"—underwent a metamorphosis that transformed it from a creature that walked by itself and lived where it chose into a subspecies, *F. s. catus,* the "clever cat of the woods." (I wonder: is it simply coincidence that two well-known cartoon cats are called Felix and Sylvester?) *Catus* can also be translated as "prudent" and "circumspect." *F. sylvestris*, in at least fifteen subspecies, all shorthaired and striped, can still be found in North Africa, Europe, and Asia. It was the North African version, *F. s. libyca*, that the Egyptians transformed. Some people classify the domestic cat simply as *F. catus*, as if it had been made completely independent of its wild ancestor, but that is misleading, for wildcat and pussycat can interbreed. There is a difference, however, between a wildcat and a feral domestic cat. Some of the Web sites featuring pictures of wildcats show not just the tiger cat of the woods but cats of many colors, with long hair or short. The noses of some of these so-called wildcats are pushed in, a feature completely absent in their woods-cat ancestors. But these are not depussified representatives of *F. sylvestris*. They are *F. s. catus*, plain and simple, and their not-so-distant predecessors once purred under the stroking hand of a human being. And we're still messing around with *F. s. catus*, turning out cats like the Sphynx, which looks naked and wrinkled but is actually covered with a fine coat of down; the La Perm and the Selkirk Rex with curly fur; long- and shorthaired Scottish Folds, the ears of which are bent down; the spotted Ocicat, cobbled together from the Siamese, Abyssinian, and American Shorthair; and the Bombay, an all-black breed with huge copper eyes that was created in the 1950s by crossing a sable Burmese with a black American Shorthair.

It should be said that the true wildcat and the feral domestic cat do have one feature in common: healthy teeth. My daughter, a vet, participates in a Trap-Neuter-Release (TNR) program, in which people catch feral animals, bring them to her for spaying or neutering, plus shots for rabies, and then take the cats back home. She noticed that the teeth of these unsociable creatures were in shiny clean condition—no plaque and almost no tooth loss—unlike the pets with dirty, cavity-ridden snaggleteeth that visit her clinic. She says that the difference is due to diet. Mice, eaten fur, bones, and all, act like dental floss, as do the hard chitinous exoskeletons of grasshoppers and crickets. It looks as if foods for pet cats need to be reformulated.

She points out, too, that our bargain with the cat is not like the bargains that we have made with other domesticated animals: I'll feed and care for you, and you, in turn, will provide me with benefits. The cow honors its part of the bargain with milk, meat, and leather; the horse, with transportation and draft-animal work; the sheep with meat and wool; the pig, with bacon and ham, plus garbage disposal; the chicken with eggs, Sunday roasts, and much more, as shall be related later; and the goose with Christmas dinner and a gaggle of nursery rhymes. The dog in its many incarnations is a jack of all trades, pulling carts, guarding people and places, herding sheep and cattle, retrieving game, hunting deer and bear, sniffing out explosives and drugs, catching rats, toting casks of brandy to stranded travelers, and much more. But the cat? Why worship the domestic cat? Because the cat seems an intermediary between the mortal and the divine. Because the cat kills the rodent that spoils the grain that could have fed the people. Because the cat appeals to our eyes and hands. Because the cat purrs. Because, because.

*

Purring, that deep-throated rumble, may be almost inaudible or may be heard two rooms away. Why do cats purr? How do they make that seductive sound? The second question is more amenable to an answer. Most likely, this audible vibration comes from a nerve-activated back-and-forth play between the muscles of the pharynx and the diaphragm. Its frequency lies between 25 and 150 hertz. In domestic cats and the other purring members of the family Felidae, the Cat family, the purr continues uninterrupted as the animal breaths in and out. All of the smaller felids purr, including the lynx, bobcat, ocelot, puma, Africa's leopard-spotted serval, and the tuft-eared caracal of Africa and western Asia. Of the larger cats, the cheetah and the snow leopard set up a rumble that occurs as they inhale and exhale. Zookeepers, circus people, and researchers who work with the larger cats, like lions and tigers, claim that they also purr, but that the sound comes forth only on exhalation. The big cats do something that the smaller cats cannot: they roar. But why should a smaller cat roar? They clearly exult in a sound that's more in keeping with their size.

The reasons that cats purr remain elusive, although the puzzle pieces begin to fit together. In our usual self-involved way, we like to think that a purr means happiness, that it's a positive response to being stroked and signals pure pleasure in the attention with which we worship a pet cat. But the sound is not elicited solely by contact with us. A cat purrs while making bread, giving birth, and nursing kittens. In turn, the kittens purr as they pull on the teats. A cat may purr in stressful situations, like visiting the vet or incurring an injury. My Omega cat, suffering from a urinary-tract infection, purred when a vet pressed his bladder to collect a specimen of urine. A cat at death's door can purr.

It takes energy to make a purr. So what benefits accrue from such behavior? Studies suggest that the vibrating interplay between pharynx and diaphragm is central to both communication and self-healing. For communication, think of the mother and her kittens, and think of our own cats' relaxed responses to being patted. They certainly do let us know that our actions are okay with them. As for healing, that a purr may have such power is a discovery that may have important ramifications. No irrefutable proofs yet exist, but a grand body of evidence points to strong possibilities that purring enhances bone density, relieves pain, and promotes quick recovery after surgery. Some of the evidence is comparative: dogs, for example, experience far more diseases of their muscles and bones and come back from surgical procedures far more slowly than do cats.

Dogs, as we know, do not purr. Some evidence is inferential: people, especially women, are advised to exercise in ways, like walking, stair-climbing, and (hurray!) dancing, that will keep osteoporosis at bay by putting mild stress on their bones. Cats in general have strong bones, a fact that seems out of character with their propensity to sleep more hours than they spend awake. The purr may well be their way of walking, stair-climbing, and dancing. The most likely agency for all these effects is that the vibrations set up the purr's pharyngeal-diaphragmatic dynamics. Some researchers have tested this theory in novel ways, such as putting chickens for twenty minutes every day on a plate vibrating at frequencies between 20–50 hertz. And, marvelous to say, the chickens' bones grew stronger! Experiments with rabbits showed that frequencies of 25 and 50 not only increased bone strength but led to speedier healing of fractures.

These results imply that humankind might benefit nicely if there was a whole lot of shaking going on. People with osteoporosis and astronauts losing bone density all too quickly at zero gravity could use some artificial purring to restore bone mass or keep from losing it. People suffering from severe pain have found notable relief in an electric stimulator, which is a machine analogous to the purr.

A friend tells me that once upon a time, when she was—quite unlike her usual self—miserably sick, her longhaired orange tabby, Quatorze, lay beside her—quite unlike his usual self—and purred until she rose, recovered, from her bed.

Physician, heal thyself. The domestic cat does just that and may even choose to include us among those benefiting from the vibratory rumble.

*

With few exceptions, all domestic cats purr, but some cannot—can never—be tamed. A tiger tabby arrived on my doorstep more than thirty years ago and was invited inside after my son made a fervent plea that we keep her. He named her Thomasina Tigger Twitchy-Tail, and to that moniker an epithet was soon appended: Terrible. When she was picked up, she'd purr ferociously for about thirty seconds before striking out and raking the face of the person who'd only wanted to pat her. My veterinarian daughter, Lisa, tells me that this is psychotic behavior and that any cat possessed by such uncontrollable rages should be put down. At that time, though, Lisa was far from being a vet, and we did not euthanize Terrible Thomasina. She was returned to the wilderness whence she had come.

If a cat does not receive pats from a human hand and become acquainted with the sound of a human voice after it opens its eyes, an event that occurs about ten days after birth, it will, in all likelihood, never be good company. If there's been no direct contact with people during its first six to eight weeks of life, there's nothing that we can do to turn wild into tame. But what about Leroy, who was certainly a street kitten at the time that my young neighbors took him in? He is outgoing not just in regard to the great out-of-doors but also in his relationships with people. He may well have been homeless for most of his life, but it is certain that when he was a very small kitten, someone played with him enough to make him believe that people were not an inimical Other but rather a natural part of what happens to cats. As for my own cats, adopted from the local SPCA when they were ten weeks old, only Omega, the tabby, had been completely socialized. He purred and nuzzled immediately on being held. An hour after arriving at my house, he claimed me by walking around and around on my shoulders and marking my neck with the scent glands at the side of his mouth. But at the shelter, black-and-white Alpha struggled to get away. Not a purr could be coaxed from him. It took thirty-six hours for him to claim me by marking my face, at 5:30 A.M. no less, and a full year before he decided that it was all right to lie upon my lap if *he* so chose. If he'd stayed at the SPCA for another week, he might not have been salvable as a house cat. I think that I got him just in time.

Sophie is another story altogether. Not long ago, she was kidnapped. It's not the first such incident on the street. Last summer, the owners of an attractive orange tabby kitten, no more than twelve weeks old, saw it being taken into a car, the not unlikely outcome of allowing the kitten to roam free. As for Sophie, it took a few days before I realized that her regular smoke-gray peregrinations were absent. She was gone. I soon found out whence and why. A onetime neighbor's sister had spotted her, and after the sister's six-year-old child had begged to keep the cat, Sophie was whisked away. Three weeks later, she was whisked back because her indoor behavior left much to be desired. She used the litter box, then ate her feces. My daughter, the vet, tells me that to do so is not at all normal; it's downright pathological. A cat will eat aborted kittens as a matter of course, but not feces. I suspect that's the reason that the cat who has become Sophie was sent into outdoor exile in the first place. She'll probably remain a constant smoky presence, purring and seeking attention throughout the neighborhood until the feline equivalent of the grim reaper—a car, most likely—comes to harvest her.

Moral: All purrs are not the same.

TWO

A Bird of Consequence

{ A Tale of Bargains }

***P**eep-peep-peep-peep.*

"The seminar was over," my daughter Lisa says. "Everybody else had gone, taking their chicks with them. But there were these two little ones left behind, *peep-peeping* away. So I said *peep-peep* right back, and here they came. It was freezing out; so, I wrapped them up in that old blue throw I keep in the car and took them home."

Nearly two decades ago, when I visited Lisa and her family at their home in a rural corner of northwest Wisconsin, I met those chicks, along with a mixed flock of other birds. She had birds everywhere—a barred owl in the screened-in breezeway, a nearly featherless goldfinch in a bathroom, and the pair of chicks in the kitchen. The first two were rehab clients—she's a veterinarian. The other two—how to describe them?—served as active, sweetly vocal kitchen companions. She'd brought them home from that teaching seminar when they were about a week old. They'd been used to demonstrate to owners of fancy show chickens how to draw blood to test their birds for pullorum disease, which is a highly contagious diarrheal salmonellosis. Somehow the two had been left behind, and that could not be countenanced. When she reached home, she unwrapped the little birds and put them into two cages that she placed upon the island in the center of the kitchen. Why in the kitchen? Because there they were safe from Biscuit, the family's rescued greyhound, who was mighty fond of chickens as objects of the chase, and the chases were invariably fatal. Lisa named one Flicka; the name of the other is lost in the deeps of time.

Neither was much to look when I met them. They were pullets that had lost their baby down but not yet put on a full rig of feathers. Their feathers-to-be protruded like quills from their pale skin. Oh, they were sociable! Except when they were asleep, they made soft chicken music. They kept up a constant, cooing, peeping, chirping, back-and-forth conversation that included Lisa when she was in the room. And she responded, of course. In a short time, Flicka and her kitchen-mate outgrew their cages and were consigned to a fenced chicken yard, along with some hens and an immense rooster the size of a tom turkey. Sad to say, Biscuit breached the fence. End of pullets.

I am certain that during their all-too-brief stay in the chicken yard, they played just as the flock of white Leghorns kept by Becky, a North Carolina neighbor, played. Those hens spent the night in a shed but ran loose in her yard during the day. What a tumble of birds! They strutted and ran, leaped and chased. To be sure, such activity stirred up snacks in the form of insects, but those chickens were also having fun. Becky, to our consternation, also kept a sleep-interrupting rooster that crowed loudly at midnight, as well as at daybreak, noon, dusk, and every hour in between.

Chickens possess an undeniable social awareness. Not only do they talk and play, but a flock establishes a hierarchy, in which one bird, occupying the summit of social influence, bosses all the others, while another finds itself irremediably at the nadir, with the rest of the flock somewhere in the peck-and-be-pecked middle. And that bird at the bottom of the heap often lives in a state of bloody nakedness, every last feather stripped from its pimply body. That unclad bird is, to my eyes, a pitiful sight, but I suspect that it accepts its fate simply as The Way Things Are. A pecking order comes into being when members of a community are kept in close proximity. It's a form of government, like office politics, that is designed to let individuals know exactly where they fit into the scheme of things. The pecking-order difference between Them and Us—between chickens and those who work in cubicles and hang out around the coffeepot—is that we are sometimes able to work our way up or down the ladder of dominance, while the birds' status remains quo. The ways in which chickens decide who is to be top bird, bottom bird, and birds between is a mystery but may well be analogous to the ways in which we establish rankings. Getting to the top takes strength, speed, agility, and wit.

*

Where did chickens come from? From those thunder-lizards, the dinosaurs—that's where. One of the grand discoveries of the late twentieth century is that all avian species, from grebes and geese to woodpeckers, house sparrows, house wrens, and—yes—chickens, most likely evolved from dinosaurs, and not just any dinosaurs but those known as theropods, "beast-footed" creatures, which walked on their hind legs (their forelegs were tiny in comparison) and dined mostly on flesh, though some were dedicated vegans. One of the best-known theropods is *Tyrannosaurus rex*, a reptile of truly intimidating proportions, but many of the theropod dinosaurs were small, quick creatures, darting upright through the swamps and jungles of the Cretaceous Period more than 66 million years ago. Fossils show that a group of them, the coelurosaurs, the "hollow-tail lizards," had features remarkably like those that would characterize the birds of the future. And some of these little theropods survived the massive extinction of the dinosaurs that occurred at the end of the Cretaceous and the beginning of the Mesozoic.

Birdkind actually got off the ground way back in the Late Jurassic, which began some 161 million years ago. *Archaeopteryx*, or "ancient bird," is the one with which we're most familiar. Its fossils were discovered embedded in limestone in Bavaria back in the mid-1800s. Unlike modern birds, it had teeth and claws on the tips of its feathered wings, and its backbone stretched out into a tail. As far as can be discerned from the remains, its leathery body lacked feathers. About the size of a crow, it could fly but only in a clumsy, short-distance fashion (much in the manner of, yes, a chicken). Another equally ancient species, *Confuciusornis sanctus*, or "holy Confucius-bird," was discovered in China in 1994. About the size of *Archaeopteryx*, it looked a bit more like something we would recognize as a bird, for the fossil record shows that it not only wore feathers on its body but sported a horny, toothless beak. Birds found in what are now Germany and China—in those olden days, they were already showing diversity, suggesting that birdkind may well have had more than one point of origin. But these winged Jurassic creatures are not thought to be the direct ancestors of the present day's more than 9,000 avian species.

Nor are some of the lineages that evolved during the Cretaceous: the Hesperornithiformes, the "Western bird-forms," a mostly flightless group that swam (think penguins); the toothed Ichthyornithiformes, the "fishy bird-forms," that most likely fed on fish; and the Enantiornithes, the "contrary birds," that readily took to the air. (I can only guess why the taxonomist chose to call them "contrary"—perhaps because they flew while the other lineages didn't.) But by the end of the Cretaceous, evolution had transformed some of the small, darting theropods into birds that we could readily identify as water birds like geese and ducks, shorebirds, gulls, and seabirds, like petrels and the soaring albatrosses. The fossil evidence shows that most of today's bird orders had appeared in the early Oligocene, 35 million years ago. Chicken-like birds, if not the chickens that we now know, certainly clucked, gobbled, and squawked back in those times. Fowl-like birds came into existence before eagles and hawks (perhaps so that when the raptors arrived, they'd find their dinners waiting).

Two questions regarding chickens crop up persistently: Why did the chicken cross the road? And, which came first, the chicken or the egg? One reasonable answer to the second is: the dinosaur came first. And the dinosaur was, like the feathered versions produced by evolution, an egg-laying creature. As for the first question, let's get that one over with here and now before the story moves into less frivolous realms. Conventional wisdom has it that the chicken crossed the road to get to the other side. But, given the playfulness of the bird, it actually crossed the road to show the possum that it could be done.

*

When and where did modern domestic chickens originate? Perhaps as early as 8000 B.C., in the forests of India, China, and other parts of East Asia. The ancestral bird was the wild red jungle fowl, *Gallus gallus*, "cock cock," a member of the Phasianidae, the Pheasant family. And the red jungle fowl rooster is indeed a handsome bird—long golden feathers on his neck, body clad in brown and bright maroon, a distinctive white patch on either side of his head, a lush and plumy tail, and sharp spurs on the backs of his legs. His direct descendants still run and cackle in their ancestral homes, though they have been extensively interbred with domestic chickens. The reason for domesticating the bird had little to do with its edibility, though it and its eggs, as well, must have been served up not infrequently for supper. The original bargain involved in domesticating the red jungle fowl was not one that traded care for eggs, meat, and feathers. Entertainment was at the top of our agenda. Exchanges of money and goods surely took place, too, either as a purchase price for the birds as merchandise or as bets. For, the primary interest of the keepers lay in cockfighting, a bloody sport that spread rapidly through Asia, the Near East, and Africa. Chicken bones have been found in Egyptian tombs dating back to the Old Kingdom in the third millennium B.C. The birds arrived in Greece no later than the 700s B.C., and the Greeks acknowledged their origin by calling different breeds Median, Tanagrian, and Chalcidian, names that refer to places in Asia. Those breeds were said to be unsuited for laying but fit indeed for combat.

Not long thereafter, various strains of domesticated jungle fowl made their way to Rome and thence into northern Europe. Spanish conquistadors introduced the battling birds into the New World (though there is good DNA evidence that the Araucana breed found in South America is pre-Columbian and crossed the Pacific with Polynesian sailors). Today, though cockfighting is widely outlawed, gamecocks may easily be found. I've seen working birds on a farm only fifteen miles away from the Shenandoah Valley town in which I live. They're gaudy, loudmouthed, and spurred with stiletto-like spikes, made more wicked by the attachment of steel spur-covers. Gamecocks resemble their wild red ancestor more closely than do any other domestic chickens.

In Rome, the human bargain with the chicken included divination. Sacred birds were tended by a *pullarius*, a chicken-keeper, who would offer them food when their services were needed; if they gobbled it up, good fortune was on its way, but if they disdained the tidbits, cackled, beat their wings, or flew off, bad luck was imminent. The flight of the birds and their manner of pecking for insects and seeds could also be read for omens. And eggs, as shall soon be told, offered a means of ensuring the birth of a human boy-child. Considerable interest also focused naturally on fattening the birds for eating and using their eggs as food, even though Rome more than once, in fits of conservatism, passed sumptuary laws that made it illegal to plump out chickens. The laws were honored more in their flouting than in their observance.

The raising of domestic chickens is given detailed coverage in the book on agriculture written by the farmer-statesman Marcus Terentius Varro (116–27 B.C.). It mentions three kinds of fowl from which the poultry-keeper might select—barnyard, wild, and African, the last of which is actually the guinea fowl, a large, honking, squawking bird. Varro zeroes in on five points: selection and purchase, breeding, eggs, chicks, and the fattening of birds for the table. When it comes to buying a cock, he suggests that the bird chosen have a reddish comb, a beak that's short, wide, and sharp, eyes dark gray or black, neck parti-colored or golden, short lower legs, long claws, and a large tail. Above all, the cock must be *salax*, eager for sex. He recommends a method for castrating a cock that would now be considered brutal: ramming a red-hot iron into the bird's groin and then smearing potter's clay over the wounds. The operation makes it easier to fatten the bird, for he loses not just his crow but his calorie-consuming sex drive.

Today, a cock becomes a capon when his large testicles, tucked within his abdominal cavity on either side of his backbone, are surgically removed.

Varro's advice on caring for chicks includes picking lice from their heads and necks when their feathers begin to sprout and burning stag horns around chicken coops to keep snakes from entering. His prescription for fattening hens sounds like an archaic version of factory farming. After the feathers have been plucked from their wings and tail, the hens are shut up and fed such goodies as pellets of meal, flax seeds, and wheat bread moistened with mixed wine and water. After twenty days, they will be plump and delectable.

The Roman natural historian Pliny the Elder (23–79) repeats much of Varro's advice on selection and fattening. His account also refers to dwarf breeds of chickens, he speaks of poultry diseases, and he adds a mention of cockfighting in Pergamum, an ancient Greek kingdom in Asia Minor. Most curiously, he reports a way in which a woman can make sure that her pregnancy results in the birth of a son. The woman in his story is none other than Livia Drusilla (58 B.C.–A.D. 29), who became the wife of Augustus Caesar. At the time, however, she was married to Tiberius Claudius Nero (not the Nero, born in A.D. 27, who fiddled while Rome burned). Becoming pregnant, Livia was eager to bear a son, and made her magic this way: she tucked an egg between her breasts; when she needed to put it down, she handed it to a maidservant, who would cherish the egg in the same way so that it would not lack warmth. Eventually, the egg hatched, and Livia's wish came true: she bore Tiberius Claudius Nero, who was to become, as Tiberius Caesar Augustus, Rome's second emperor.

In the U.S., it was only in the late 1800s that attention came to focus on fattening chickens as a prime source of meat and eggs. Of course, the birds had run around in villages and barnyards since people had first noticed and domesticated them. But only young—and therefore tender—specimens were to be roasted, fried, or fricasseed. Old, tough chickens ended up in the stewpot. In her cookbook *The Virginia Housewife*, first published in 1824, Mary Randolph gives a recipe for "Soup of Any Kind of Old Fowl" and describes it as "the only way in which they are eatable." As you can see from her recipe, given below, she was fully as interested as Varro and Pliny in fattening the birds.

> Put the fowls in a coop and feed them moderately for a fortnight; kill one and cleanse it, cut off the legs and wings, and separate the breast from the ribs, which, together with the whole back, must be thrown away, being too gross and strong for use. Take the skin and fat from the parts cut off which are also gross. Wash the pieces nicely, and put them on the fire with about a pound of bacon, a large onion chopped small, some pepper and salt, a few blades of mace, a handful of parsley, cut up very fine, and two quarts of water, if it be a common fowl or duck—a turkey will require more water. Boil it gently for three hours, tie up a small bunch of thyme, and let it boil in it half an hour, then take it out. Thicken your soup with a large spoonful of butter rubbed into two of flour, the yelks [sic] of two eggs, and half a pint of milk. Be careful not to let it curdle in the soup.

No danger that anyone would get *E. coli* from old fowls cooked to this kind of fare-thee-well!

Nowadays, unless we raise fowl ourselves, we're not likely to get a stewing hen old and tough enough to require Mary Randolph's treatment. The broiling chickens that we buy at the grocery store are slaughtered at tender ages, mostly from six to eight weeks, and almost all of them are contaminated with bacteria. The secret of avoiding illness is, of course, to cook them thoroughly. Laying hens, on the other hand, are allowed to live for two years. But left to its own devices, plus adequate forage and shelter, a chicken will live, depending on its breed, for five to eleven years.

And, oh, the breeds that we have engineered! We have tampered with chicken genetics since prehistoric times. One study says that "chicken breeding represents one of the most remarkable examples of directed evolution." The Department of Animal Science at Oklahoma State University lists sixty-three breeds of barnyard and pet chickens in the U.S. and Europe, which fall into three modern categories: broilers for meat, layers for eggs, and the rest as ornamentals, good for nothing but being strange and/or beautiful. The list does not include varieties that exist in China, Japan, and places elsewhere in the world. Some breeds are familiar, like Rhode Island Red, Buff Orpington, Leghorn, and Plymouth Rock. Others take eccentricity to various extremes. One is the Appenzell Bearded Hen, which sports beak-surrounding feathers that look like mutton-chop whiskers. Another, the Naked Neck or Turken, wears only a small tuft of feathers near the collarbone on its otherwise denuded neck. The Crevecoeur was bred not as food but as an ornament; these heads of these birds are crested with fine black feathers that look like a helmet, and their bodies are heavily clad in black with a shimmering green iridescence. The Silky breed is also ornamental—like all chickens, they are edible, but their dark skin puts off would-be diners. Their feathers, lacking barbules, distinguish them; grown birds are as fluffy and soft as chicks. Faverolles, both roosters and hens, wear beards and sprout feathers on the backs of their legs. Some, like the Modern Game breed, are skinny, long-legged creatures that look as if they walk on stilts, while others, like the Jersey Giant, are extravagantly plumped out. We've had as fine a time messing around with *G. gallus* as we've had creating weird and wonderful variations on dogs and cats.

*

As it happens, chickens turn out to be good for much more than food, feathers, entertainment, and divination. They have long served as laboratory animals for the study of embryology. They have also figured in studies of viruses, cancer, and immune systems. Then, in 2004, the International Chicken Genome Sequencing Consortium reported that it had made a draft sequence and comparative analysis of the red jungle fowl's genome. The first genome of a non-mammalian vertebrate ever to have been sequenced, it has since reached final form. Not only is the chicken the first bird to be sequenced, but it is also is the first agricultural animal and the first descendant of the dinosaurs to have its base pairs of sequence lined up. And, oh my, the genome has about a billion base pairs and twenty to twenty-three thousand genes. (DNA contains molecules of adenine, thymine, guanine, and cyclosine. The first two form their own pairs, as do the latter two, and in these pairs—AT and GC—the genetic information that makes a person a person and a chicken a chicken is encoded.) One difference between the chicken genome and those of the mammals—rats, mice, dogs, chimpanzees, and human beings—that have been sequenced is that the chicken genome checks in at about a third of the size of the others.

What good is this knowledge? And who cares? Chicken breeders, for one; with such information, desirable traits, like big meaty breasts and abundant egg production, can be optimized. Transgenic chickens—chickens with genes from other organisms, including some human genes—have already been created in laboratories; the aims are agricultural and medical, notably projects that seek to eliminate avian influenza or reduce its incidence. But more than acting as an aid to artful breeding and medical innovation, the chicken genome offers a fine tool for the study of human evolution. As Jeremy Schmutz and Jane Grimwood put it in *Nature* magazine:

> Now that the human genome sequence is essentially finished, researchers would like to do more than just identify the sequences that are translated into proteins. They also want to understand all of the regulatory structures present in a genome—structures that might, for example, adjust

> the amount of protein manufactured from a particular gene. These structures are collectively known as functional elements, and the chicken, having diverged from humans more than 310 million years ago, is considered the best example of an "outgroup" with which to identify them. Because enough differences between the human and chicken sequences have accumulated over this period, one can zero in on the precise base pairs that evolution has left alone for all these years.

Nature is not wasteful. Once successful strategies have come into being, they are not thrown out in favor of newer models. Strange and marvelous to say, the chicken and the human being share a host of conserved elements—a host that amounts to some 70 million base pairs of sequences. The chicken genome will shed light on vertebrate evolution by adding details that were not available from the sequences of mammals. The consortium that unraveled the chicken genome says, "For nearly every aspect of biology, it allows us to distinguish features of mammalian biology that are derived or ancient, and it reveals examples of mammalian innovation and adaptation."

*

Nor do the bargains that we make with chickens end with the insights that they provide into evolution's quirks and quiddities. No, indeed! Because of them my garden flourishes. The tomatoes resemble large shrubs, each butternut squash weighs at least three pounds, the beanstalks could support Jack. The reason for such burgeoning is that the vegetables have received a liberal application of chicken litter. An acquaintance and her husband raise chickens for Tyson, and every eight weeks or so the population in the coop turns over as old chickens are sent to market, and new ones take their place. Interim, a mountain of chicken manure needs to be removed, much of it seasoned and dry enough so that it has no reek. I borrowed a truck, filled it with litter, and spent the afternoon literally shoveling shit. Wheelbarrow-load by wheelbarrow-load, it was distributed over the gardens in the back and front yards. A friend helped me till it in. Whee! Chickens are a splendid source of fertilizer.

I dream of having a chicken coop in my backyard. In this imaginary coop, I raise Rhode Island Reds, the eggs of which are a soft, warm brown, and the yolks of the eggs a rich, deep yellow because the birds have not been fed on commercial chicken mash but have scratched for seeds and insects and rooted through an open compost bin. Even though the town has laws that make the keeping of livestock and poultry illegal, I have heard roosters crow within its limits. Some years ago, one neighbor even kept a grass-mowing goat in his yard. I wonder if I could con the preservation commission, which passes judgment on all matters pertaining to the town's five historic districts, into allowing me to house a small flock of Rhode Island Reds in the Newtown District. The rules of the preservation commission are few but strict: If it was once here, then it may be restored, and, if it's new—a shed, perhaps, or a greenhouse—people must not be able to see it from the street. So, a greenhouse is out of the question for me. I do know from meeting the tiny old lady whose father built my house in 1910 that horses were housed in a livery stable on the alley behind the house, and that her mother kept two milk cows. I would think that chickens must have been part and parcel of the neighborhood, as well. Then, I remember Becky's rooster that crowed around the clock. It was more irritating than a yap-yapping dog or someone grinding the starter on a car, and more than once, my husband threatened to turn that bird into cockadoodle-stew. End of chicken dreams.

*

I remain an ardent chicken-watcher, nor am I reluctant to consume them. With all the services that *G. gallus* has provided for us in the past, and shall continue to provide now and in time to come, clearly the chicken is not just a bird that has been sequenced but a bird of consequence, an important species without which our world would be quieter but far less colorful and tasty.

I make my own bargain with the chicken: I shall honor and savor you because you have given me not just food, fertilizer, entertainment, and insight but also stories.

THREE

Whitetails

{ *A Story About Consequences* }

The last week in June: a sky as blue as flax flowers overhead; on all sides, a green in fields, on trees that has not lost its freshness to the sultry summer heat; before us, the last seven miles of the winding country road that takes my husband, the Chief, and me home. We've come from town, twenty-three miles away, where he had an appointment with a doctor at the Naval Hospital, a place—no, a limbo—that always involves a wait that can't be calculated in advance, though it's invariably long. We talked, I read a little, and the Chief fidgeted. Both of us were all-fired eager to escape the tedium, to return to the sweet dailiness of life on the banks of the wide and salty river Neuse.

THUNK!

Shock. Can't move. Can't think.

Motion and reason return an eon later. I lift my head, stop the car, take my hands off the steering wheel. The Chief asks if I'm all right. Yes, as much can be. Then he's out there on the road dragging the deer to the grassy verge. Blood covers his hands. And I see that it's not just one whitetail but three that I have killed. The hooves of twin fawns protrude from the doe's burst belly.

An astonishing thing happens. The man who had been mowing the churchyard with a lawn tractor comes quickly to us and calls the state troopers on his cell phone. It's our neighbor, Del, who has a landscaping and lawn-care business. Then he kneels over the doe, pulls one of the fawns all the way into the air, and administers mouth-to-mouth resuscitation. I have never before seen a human being trying to resuscitate a wild animal. It can't happen often. But it's too late. The doe and her twins are gone to whatever realm receives their innocence, a realm in which they are made whole again, safe from guns and cars, to browse freely and forever.

The church is open, and the Chief is able to wash his hands. We sit, then, under the roof of the outdoor pavilion and wait for the trooper to arrive. Sometimes, if response is quick, the person who has hit and killed a deer

may be given a permit to possess the deer; then the animal may legitimately be taken home and dressed out. Sometimes, in the boonies, passersby don't bother about a permit but just load it into the back of a pickup truck. But no one stops by, and it takes the trooper an hour to make his way out here. By that time, the deer is beyond salvage, for the ninety-degree temperature has speeded the process of decomposition. My car is still drivable, though sadly smashed; after making out an accident report, we continue our journey home. The car sits in the yard until a rollback can tow it to town for three thousand dollars worth of repairs. Six months later, the car will be totaled by a Subaru running a red light, but that is another story altogether.

Once home, we remember other encounters between local vehicles and deer. One night, our neighbor Bonnie hit a button buck, a male fawn, that is, with the nubs of antlers emerging from his head. She stopped her car and put the unconscious animal in the back seat. He came to with a great flailing of hooves just as she was pulling into her driveway. Fortunately—or unfortunately, depending on the point of view—she and her husband were able to subdue him. The next morning, their freezer held a brand-new stash of the most tender venison.

I recall the rule taught to big-rig truckers: *Hit the deer. Go ahead and hit it. If you swerve, you risk running off the road. You risk your own life, not just that of an animal.* I once received a call in the wee hours from my younger daughter, who had spent a year as a trucker. Her voice was shaking. "I did it," she said. "I drove right through a big buck." No, she wasn't hurt, though the truck's dented cab still dripped with blood. Even though she had done the right thing, the suddenness of the encounter and its lethal consequences had put her into a state of shock. And night's darkness magnifies distress. She had called the police to report the collision. I advised that she go to bed and do her best to sleep it off, that the arrival of daylight—and the necessity for making a report to the trucking company—would put the accident into perspective.

Car or truck smashing into a deer, such Minus-Minus accidents in which neither party wins are almost as frequent as sunrise and sunset; they are to be expected; and they have warranted considerable study by state and regional governments. An unpronounceable, dry-as-dust acronym has been coined to describe these encounters: DVC, which stands for Deer-Vehicle Collision. According to the United States Department of Agriculture's Wildlife Services, DVCs cause more than 200 human deaths and almost 30,000 injuries annually, with the bill to repair damages amounting to more that $1 billion. And, because the whitetail and human populations increase every year, along with the miles that people drive, the number of accidents also mounts. In my home state, Virginia, a report issued in 2006 estimates that DVCs have increased at least ten-fold since 1966. The agencies that deal with wildlife and traffic now invest much effort into preventing these accidents. They range from putting up deer-crossing signs and clearing foliage from roadsides to building highway underpasses and overpasses that afford animals safe crossings. Motorists are duly cautioned on how to avoid DVCs: drive defensively, do not speed, do not swerve to avoid hitting a deer, be especially alert at dawn and dusk and during the fall mating season. Mysteriously, the report fails to include another necessary injunction: do not talk on a cell phone while behind the wheel. Commonsensical advice—without even thinking, we should be practicing it when we drive. Yet, all the caution in the world cannot prevent galvanic encounters between flesh and metal. *THUNK!*

*

White-tailed deer and their kin—more than thirty species worldwide—have long garnered more than a share of reverence. They are the quintessential wild animal, as their English common name implies. It comes from the Old English *deor*, a word closely related to the German *Tier*, both of which refer to any never-domesticated mammal. Today, *deer* evokes mind-pictures of a long-legged creature that is superlatively elegant in its grace, speed, and beauty. Its coat shines bright cinnamon-brown in summer, gray-brown in winter. White circles its eyes and nose and decorates its chin, belly, and the underside of its tail. A raised tail may flag an alarm or, with a doe, give follow-me signals to her dappled fawns.

Deer are the subject, too, of annual rituals of cunning adoration—the autumnal hunts. Would-be deer slayers take to the field, to the woods and mountains, in crisp fall weather; wait in as perfect stillness as a human being can achieve; and hope that a clean shot will crown their day. Nor do the rituals end with the kill but continue. The hunter brings the deer to a gambrel and hangs it high. He—or, with increasing frequency these days, she—skins it, removing the stomach and entrails with great care, and cutting the meat—hams, loins, backstrap—into pieces suitable for packaging. Then the carcass is buried, the blood washed off human hands. And still the rituals go on, for the stories of the hunt demand telling. The telling is a form of thanks for the animal's grace and its meat, for clean air and the woods' silence, and for the companionship of other hunters.

So, I am sure, has every true hunt proceeded since a human being first set out to bring down a deer. Deer are talismanic animals, variously signifying desirable traits like nimbleness, wisdom, and fertility. They leap in pictographs painted on the walls of French caves and in petroglyphs pecked or scratched into rocks as close to home as Arizona, as far afield as Tibet; the images constitute visual prayers for success in the hunt. And deer may have been considered guides to the underworld. Stags wrought of gold with huge, intricately curled racks of antlers have been found in Scythian burial mounds more than two millennia old near the village of Filippovka on the southern Russian steppes. And Scythia is the scene for a wintertime deer hunt described by the Roman poet Virgil in 30 B.C. Snow fills the sky and lies heavy on the earth:

> . . . the deer, crowded into compact herds, are numbed by the
> snow's weight, with just the tips of their horns to be seen.

> These the Scythians do not hunt with dogs released or strung-out nets or the scarlet feather meant to scare them into running, but, iron blade in hand, they strike the deer trying in vain to breast the mountain of snow, and kill them as they bleat loudly and carry them off with a fierce cry of joy.

Joy, yes, for in a season of desperate cold, the hunters have obtained fresh food and an extended lease on life for themselves and their families. And surely they enfolded their success into the necessary stories.

Deer make other notable appearances in the wilds of ancient literature. In his hymn to Artemis, patron deity of virgins and of the hunt, the Greek poet Callimachus (flourished about 270 B.C.) writes of the goddess hunting in the mountains with her seven Spartan bitches running beside her. She spots five does "bigger than bulls, and their horns shone with gold." Likely they are red deer, one of the world's largest species, and they have been emblazoned not only with antlers of precious metal but also hooves of brass, sun-burnished signs of divine favor. Light shoots off the antlers like sparks. The does gleam with enticing brightness against the greens and browns of the woods, the dull grays of the rocks. Forthwith, Artemis catches them. One escapes, but her four sisters are bridled with gold and hitched to the goddess's golden chariot.

The tale of the fifth doe is also told to this day. She remains free for a time. But the third labor set for the Greek hero Heracles is the capture of the so-called golden hind, the great doe that had eluded the goddess. The story of his hunt comes to us in several versions, and I shall tell the one that seems truest to the patience, skill, and strength that hunting entails. The hind is sacred to the goddess Artemis, and Heracles knows that deity is not to be trifled with and that he must not kill her animal lest vengeance be exacted. So, for a full year, he stalks her as, fully aware of her pursuer, she roams her browsing grounds to the east and to the west and as far north as the Arctic. Finally, weary of the chase, she takes refuge in Arcadia, a mountainous region in the central Peloponnese. There Heracles shoots her with a single arrow that does not draw blood but simply pins her forelegs together so that she cannot run. Then he picks her up, slings her over his shoulder, and walks mile after mile to Mycenae, where he at last lays her down at the feet of the king who commanded her capture.

Some hunts do not engender tales, for the hunters only parody the rituals. Once upon a November, I saw six men roaring on three motorcycles down a fire road in the Appalachians—*vroom, vroooom*. The noise would have warned off any deer within hearing range. The men were properly outfitted in camouflage, and their guns were strapped to their backs, hands being needed either to hold onto handlebars or to hang tightly to the person driving down the bumpy trail. But hunting was probably never the point of the expedition. More likely, it was meant as an escape from dailiness and an excuse to tie one on. I can't castigate them. But there is one kind of hunter who should be strung up on a gambrel and skinned alive—the scofflaw who hunts for the kill but not for the meat and so leaves carcasses to rot where they may.

❋

Today it is hard to imagine that white-tailed deer were ever scarce in North America. The flash of their white tails, lifted aloft as they ran, was a frequent sight in colonial days. When Jamestown was settled in 1607, the species flourished, not just in Virginia but throughout eastern North America. But by the early 1900s, not much more than one hundred years ago, the population had been reduced to a paltry half-million because of habitat loss and unregulated hunting for meat and hides. Recovery began in the 1930s, and today the animal may be found in every state, except Hawaii and Alaska, as well as in southern Canada, throughout Central America, and in South America as far below the equator as Bolivia. In fact, we suffer a surfeit of deer, and all our strategies—guns, fences, loud noises, repellents—have not kept them in check. They travel our highways and find browse in our cities and towns. Our gardens are devastated, our cars wrecked. For some of us, the talisman has become the enemy.

And just what is the whitetail? Three points of view coexist uneasily. Some people see the animal as worthy prey that provides the heady excitement of the chase and also nourishing food for the belly; others see it as a destructive pest; and yet others as Bambi, sweet and innocent, its life always to be spared. Technically, the whitetail is one of five North American members of the Cervidae, the Deer family, the name of which comes straight from the Latin and is the word that Virgil used to name the animal hunted by the Scythians—*cervus*, "deer." Scientifically, it's *Odocoileus virginianus*, the "hollow-tooth of Virginia." Although the genus name might indicate that the teeth are empty inside, it actually refers to the hollow, like a miniature mountain valley, in the molars and premolars—all the better to chew with. The name's inclusion of Virginia means that the type specimen was taken and described in that colony. A close cousin that shares the same genus is the mule deer, *O. hemionus*, "hollow-tooth mule," which ranges the western half of the United States; its huge, upstanding ears look like those of the horse-donkey hybrid for which it's named. The third species is the far northern caribou *Rangifer tarandus*, "reindeer reindeer," a reduplication that combines the Latin and Greek names for the animal. The deer known variously as caribou and reindeer are the same species; the distinction between them is that the former name applies to those that live in the wild, while the latter name designates those that have been domesticated. *R. tarandus* is the only species of deer in which both sexes grow antlers annually; the females can boast racks as glorious as those of any male. The fourth cervid species is the world's second largest deer, the North American elk, *Cervus canadensis*, "deer of Canada"; it's also known as the wapiti, a Shawnee word meaning "white rump." The moose, the

largest of all the world's deer, is the fifth cervid found in North American; formally, it is known as *Alces alces*, "elk elk," which has an excited, exclamatory ring and, indeed, their common name in Europe is "elk." They range across a northern tier of both North American and Europe, where they may be found from Scandinavia to Siberia. Their New World name, moose, comes from several Algonquian languages and means "twig-eater."

All five—all deer, in fact—are ruminants like cows, llamas, giraffes, and others; gobbling their food, they send it to the first two of their four stomachs, where it separates into solids and liquids. Then, when they have a moment to relax, they regurgitate this cud and chew it at their leisure—a strategy well suited to a creature as skittish as a deer. And all deer belong to the Order Artiodactyla, the even-toed ungulates, along with pigs and camels. The even-toed animals support their weight equally on their third and fifth toes rather than primarily on the fifth toe, as do odd-toed ungulates like horses. The word "ungulate" comes from the Latin *ungula*, "hoof." What are these hooves made of? Keratin, the same material that comprises human toenails, except that it's far, far thicker. (Imagine trying to walk upon your toenails!)

The populations of these five Cervidae all underwent nearly catastrophic declines with the advent of settlers. Caribou have almost been extirpated from Western states except for Alaska, where they are the prime targets for hunters. Western Canada also supports considerable herds. Their numbers undergo fluctuations that are natural rather than caused by humankind. Moose populations are currently stable in the U.S., while those of elk are increasing. Elk, which was once the widest ranging of all North American deer, is now being reintroduced into some of its former territory. New herds have been established, for example, in Wisconsin, Pennsylvania, Kentucky, and Virginia.

Virginia's elk population wandered into the state through its porous border with Kentucky. Because the animals can hardly be sent home, Virginia now offers a hunting season for the species. The numbers of mule deer are also on the rise, bringing both economic gains and losses, including a goodly share of *THUNK*s. But mule deer do not have the countrywide range of whitetails. Many places in the nation have whitetail populations that press the limits of their habitats to sustain them.

Whitetails are creatures of the edge, the transitional interface between woods and fields. Both venues provide browse, and the woods offer shelter: a place to rest and chew cud, a place to give birth. But the animal is supremely adaptable. With more and more of its accustomed habitat yielding to housing developments, it has moved into the suburbs. If there is the slightest suggestion of shelter—wooded lots in the neighborhood or a park fringed by trees, the whitetails are made to feel welcome. And the people whose yards are frequented by deer are often admiring—until the backyard vegetable garden is decimated and flowers along with foliage nipped off the azaleas. A friend who lives in the Virginia Piedmont finds herself on an interface between actively enjoying deer and damning them for the ways in which their appetites affect her garden. Sometimes, she counts as many as a dozen deer in her capacious, tree-surrounded yard. She finds much pleasure in watching wildlife from a kitchen or living-room window, but when it comes to her herb garden, she sighs, puts makeshift chicken-wire fences around her basil and oregano, and hopes for the best. And she tries to plant flowers and shrubs, like thorny pyrecantha, that whitetails find unpalatable. Hunger, however, will lead a deer to try almost anything.

The economic bounty that whitetails bring in the form of license fees, hunting and hunting supplies, meat, and recreational watching is still greater than the dollar-value of the damage that they cause to vehicles, crops, ornamental plantings, and the ecology of their habitats. They can change the biodiversity of forests and fields by eating only what they like best and sometimes eating it into local extinction. They also act as hosts for the blacklegged tick that is a vector for Lyme disease. Their increased numbers make for an increase in the diseases from which their herds may suffer—bovine tuberculosis and chronic wasting disease (CWD). The latter is a transmissible spongiform encephalopathy, as are mad-cow disease, sheep scrapie, and Creutzfeldt-Jakob disease in human beings. A fatal disease that attacks an animal's central nervous system and kills slowly, CWD is found in herds that inhabit several Canadian provinces, eight states west of the Mississippi, Wisconsin, and, in 2005, West Virginia. Luckily, infection is not widespread. Herd reduction is the means used to try to control the incidence of the disease. But such management will not suffice if deer populations continue to grow. Given optimum nutrients, deer can double their numbers within two years. Is it safe to eat the meat of infected deer? Science has not yet issued a definitive answer.

Yet, how we admire them! The white-tailed deer has been named as the state animal of Arkansas, Illinois, Michigan, Mississippi, Nebraska, Ohio, Oklahoma, Pennsylvania, and South Carolina; Wisconsin has named the white-tailed deer as the state's wild animal, with the badger as its official animal and the dairy cow (what else for the state of cheese heads?) as the domesticated animal. Because we find them beautiful—and beautiful they are, indeed—we tend to encourage their presence near our homes. The suburbs offer them many incentives for moving in. Prime among them is freedom from hunters and other predators. Deer also find in our yards a variety of succulent vegetation greater than that offered by the wildlings in the woods. Then, because we want them to thrive, we feed them.

There are two ways for deer-lovers to go about feeding them, one benign but labor-intensive, the other deleterious. The first consists of creating a food plot of at least an acre to attract the animals. The Virginia Department of Game and Inland Fisheries (DGIF) issues this caveat: "Food plots should never be used in a manner that will draw deer into conflict with people or other animals. They should be established only in rural settings away from residences and roads." So, the food plot is an enterprise for hunters and hikers rather than suburbanites. With the goal of attracting deer for hunting or viewing, it involves site selection, soil tests, and site preparation before the seeds can be sown. Consideration must also be given to choosing a succession of plants that will provide forage in every season: for cool weather, milo, wheat, and turnips; for the hunting season, rye and wheat; for warm weather, cowpeas, sunflowers, alfalfa, and various clovers. If the plot is successful, it will summon not only deer, but wild turkeys, pheasants, quail, songbirds, and small mammals, including groundhogs. In any event, it does not take the place of natural habitat, where deer dine on buds, bark, and twigs. Proper nutrition for whitetails comprises four main elements: green forage, like clover; grain, such as waste grain in fields; soft and hard mast, including persimmons, dogwood berries, and acorns; and woody browse. A food plot serves only as a supplement. It is not considered baiting—an illegal strewing of corn and pellets only during the hunting season, and whitetails may be hunted and taken in food plots.

The other kind of deer-feeding is illustrated by a story told to me by Sally Logan, a friend who lives in Chapel Hill, North Carolina, on a two-and-a half-acre lot. The property boasts many trees—ironwood, sourwood, sweet gum, yellow poplar, red and white oak, and pine, and it's located near a wooded area used by Duke University for forestry and environmental studies. Ideal deer habitat! Some of the neighbors have tall deer fences. But, she says, "When I moved here in 2001, I decided to live with the deer rather than keep them out." From the beginning she saw deer at a distance—"legs and a flash of white tail in the woods behind my house." After several months, they became bolder, and Sally began to concentrate on landscaping with plants that they did not eat. If a salvia remained in place without a nibble, she bought more. As for tomatoes, she fenced them in but bought other vegetables at a farmers' market. And all seemed natural, live-and-let-live, until the summer of 2004 when drought struck so hard that local creeks dried up. Sally noticed that the water in the birdbath was gone every morning. It wasn't long before she saw a doe coming in to drink, then returning to the woods to nurse twin fawns. Sally says, "Since the weeds were dying in the woods and even the trees were drooping, I put some shelled corn out for her. By fall, she and her two fawns, both button bucks, were joined by a spike buck, who, I assume, was her offspring from the previous year."

As Sally knows, male offspring leave the mother in the fall of the year that they are born, while females may remain with her for as long as a year. The foursome represented a highly unusual family grouping. Sally continued to feed them and spread corn in another part of the yard as well, so that other does could find food. Sally says that if a doe approached others as they fed, she would walk toward them with her head low and gain admittance by touching noses with a member of the group. When they had finished feeding, they would groom one another, removing blood-swollen ticks from ears and the region around the eyes. But during the fall rutting season, the four had the corn all to themselves, for the other does did not venture close to the young buck. Sally says, "The next spring the doe was fat and pregnant. I could see her udder filling. And I wondered if the family would stay together." The doe left, and for some weeks, the young buck chased other does away. He and the nearly grown younger bucks continued to eat the corn that Sally spread. Soon, however, the other deer became more venturesome and moved in. She continues, "The small family's mother was gone for about six or seven weeks. When she returned, she looked terrible. Her ears were raw from bites; her head was well-nigh covered with bloated ticks. Sadly, her udder was no longer full. I never saw a fawn." The other deer prevented her from eating with them by chasing her off, even though Sally put out two batches of corn every morning. She has surmised that the doe's pitiful condition on return was due to the fact that the spike buck drove away the does that might otherwise have groomed her.

Eventually, the herd being fed in her yard increased to twenty. "I made the mistake," Sally says, "of feeding some of them in the afternoon instead of only in the morning, and they were staying in or near my yard all day long, nibbling on plants they hadn't touched before." She stopped feeding them in August 2006. But a Nor'easter just before Thanksgiving gave them some new provender: it blew down Sally's largest oak tree, which afforded supremely tasty browse.

This story is exceeding strange. The strangest part of all is the failure of the bucks to remove themselves from the maternal orbit. While Sally has attended scrupulously to observing the behavior of the herd that frequented her yard, it seems meet to check out events with a wildlife biologist. I speak with Nelson Lafon, an expert employed by the Virginia DGIF and a specialist in the study of whitetails. "Young bucks," he tells me, "are full of vinegar." In his opinion, the spike buck was not the yearling offspring of the doe but rather her mate. Opportunity likely presented itself to him in several ways. First, he found an assured food supply and stayed around to take advantage of it. More important, he was able to do so because no bucks more mature than he hung out in his neighborhood. He had no competition. The dependable larder in Sally's yard is also the probable reason that, instead of taking off before the fall rutting season began, the two button bucks stayed with their mother long enough for their buttons to become full-fledged spikes, which they shed in January 2005. As for the doe's torn, tick-infested appearance on return to Sally's yard, Nelson Lafon says that her bedragglement was not necessarily due to lack of grooming; in some years, tick populations undergo a great surge, while in others, they crash almost into invisibility.

Thinking that providing food will enhance the animals' survival or stimulate the growth of antlers, many people set out corn, pellets, and saltlicks for deer. But Sally's story has a moral: To feed deer can lead to an unnatural disruption of their inborn habits. Nor is that the only reason to forego setting out fodder. It is unlawful to set out food that attracts deer during the hunting season and its onset. The only legal means of feeding wildlife during these months is by establishing a food plot. Nor do the reasons stop there. The Virginia DGIF sets them out succinctly at its Web site, and they apply nationwide to any cervid species:

- Feeding can cause wildlife to lose fear of human beings, a situation that can be dangerous both to people and to animals.
- When many deer use a feeding area, the possibility of spreading disease among the animals is increased. Attracting deer to feeding sites has been linked in some states to the spread of tuberculosis and chronic wasting disease.
- Feeding can actually harm deer. In winter, a deer is well adapted to survive by eating the foods provided by Mother Nature. A rapid change in a deer's diet can leave the animal unable to digest the new food and can have a deadly outcome.
- Corn and other feeds are sometimes contaminated with aflatoxin, a substance produced by several related fungi. It can poison deer and other wildlife.
- If feeding deer does enhance their survival, the number of deer may increase beyond what the local habitat can support.

And the DGIF urges us to keep the "wild" in wildlife. The message is plain: Do not feed the deer. No matter how beautiful they are to human eyes, no matter how they may seem to be suffering, we have not been granted any right whatsoever to upset the balance of nature so that a personal sense of pleasure or duty may be served. Whitetails are superabundant, and so are people. The situation baldly illustrates a Them-and-Us entanglement that won't go away. It might seem to be a Plus-Plus arrangement in which both we and the deer profit: we gain ample opportunities to watch, and the deer benefit from the food that we provide. In reality, doling out corn and other goodies creates a Minus-Minus symbiosis, for we flout the natural course of the animals' lives, and they become dependent to the point that they flout their instincts. But we do have the ability to manage our encounters with deerkind and to work for a Plus-Plus outcome, in which our needs and those of the whitetails are both addressed.

What do whitetails need? The same things that we do: food, water, shelter, and plenty of sex. They need habitats that can sustain their numbers and numbers that do not exceed the limits of existing habitats. What do people need from whitetails? Abundant watching and hunting opportunities, but fewer collisions and less damage to forests, crops, and gardens. Our responses to whitetails run the gamut from heartfelt adulation to outright antipathy. Some of us focus on individual animals, while others see the species. Because the distance between the animal activists and the romantic Bambi-ites on one hand and the hunters and wildlife professionals on the other is as great as the distance between the North and the South Poles, forming management strategies is a daunting task—damned if you do and damned if you don't.

Oh, but we are ingenious! We've come up with all sorts of tactics and devices to minimize the conflicts between deer and people. Some have already been mentioned—fences, repellents, loud noises, and putting in plants that deer usually do not deign to eat. One noncommercial repellent that relies on the aversion of deer to the smell of urine has come to my attention: placing clumps from a cat's litter box around plants that you wish to save. Fencing is the most successful way to Bambi at bay. It can surround a property or simply enclose, say, the tomato patch or a single fruit tree. But take note: erecting fences to keep deer in rather than out is not allowed. Whitetails can also be diverted from an area in which they have become a nuisance by setting up supplemental feeding stations a mile away. But this expensive tactic could lead to an increase in the deer population. Dogs confined by invisible fences can be used effectively to keep deer out of orchards. The options for reducing DVCs are abundant, ranging from signage to roadside reflectors to warning whistles that are attached to cars. Reflectors—I've seen those narrow rectangles, mounted a foot or so off the ground on posts, along Interstate 64 as it crosses the Blue Ridge. According to a handbook on managing Them-and-Us encounters, "Reflectors deflect the headlights of passing cars, creating a wall of light that shines parallel to the road and thus, possibly, discourage the approach of deer." The significant word in this description is "possibly." No hard evidence has proved their effectiveness. As for the whistles, the sales pitch that lures us states that they employ ultrasonic sound to frighten away the deer; the problem here is that the hearing range of deer is like ours—if we can't hear it, neither can they.

We've also come up with non-lethal ways of reducing deer populations. The animals can be lured to entrapping nets and, after capture, relocated elsewhere. Then, various means of controlling their fertility have been studied for the last forty years. They range from surgical sterilization to delivering anti-fertility drugs by darts and the so-called "bio-bullets," which are biodegradable projectiles that can also deliver vaccines and antibiotics. These ways of keeping populations in check should appeal to people who condemn hunting, but, alas, they are often understood as interference with the plans of all-knowing Mother Nature.

Still, hunting currently provides the best tool for managing deer populations. But hunting—any kind of hunting, be it with guns or bows—still raises the hackles of many people. I have seen a photo of activists holding signs that read STOP BOWBARISM. By now, it should be clear that I am of the opposite persuasion. Hunting seems to me not barbarous at all but rather a beneficial way of maintaining herd levels so that the animals do not succumb to the stresses of trying to survive under overcrowded conditions. In the matter of bows, a skilled marksman can use them well and silently to thin herds roaming the nervous suburbs. Better that deer die from a swift, clean shot from gun or bow than from rampant starvation and disease. It occurs to me to wonder, nevertheless, if the protesters—these people who hunt the hunters—have rituals and if their gatherings result in stories.

Humankind has always hunted. Nowadays, throughout the U.S., the hunts are strictly regulated, with limits set on the game—be it deer, bear, birds—allowed each hunter. Many of us have left the hunt behind, and some of us, eschewing even domestic animals, have abandoned the carnivory for which our dentition shows we are designed. That choice is to be respected. It may help the protesters to respect my choice when they hear that hunters often donate their prey to food banks. Of the fifty states, forty-six are served by organizations such as Hunters for the Hungry, Farmers and Hunters Feeding the Hungry, and Sportsmen Against Hunger. The program in Iowa is called Help Us Stop Hunger (HUSH); its aim is to turn too many deer into provender for people with too little food. Soup kitchens there are now serving up venison chili, venison meatloaf, and even venison stroganoff. One Iowa hunter describes the situation as "win-win. I'm out in nature as a sportsman, and the hungry get the venison." And what prime meat it is—far leaner than any beef, with a surpassingly mild flavor. I've heard complaints about its gaminess but suspect that any off-taste has to do with how the meat is prepared, not with the animal from which it comes.

To my own great good fortune, my freezer is full of venison—backstrap, haunch, burger—put there by my young friend Bill Funk, who hunts with his father and does not have enough room for this bounty in his own refrigerator. As the coda to this story, I offer the recipe for the meat my family and I ate on Christmas night 2006. The instructions were adapted by my elder son, Peter, from those given on the roasting-bag box. We accompanied the meal with long-grain and wild rice combined with seasonings, homegrown broccoli, and butternut squash pie.

Venison in Oven-Roasting Bag

Ingredients

1 turkey-sized oven-roasting bag
1 tablespoon flour
¼ cup water
1 teaspoon thyme
½ teaspoon salt
½ teaspoon pepper, freshly ground
1 10- to 12-pound deer ham
2 or 3 large onions, peeled and cut into chunks
½ pound baby carrots

Preheat the oven to 325° Fahrenheit.

Place the flour, water, thyme, salt, and pepper in the roasting bag. Shake it to coat the interior of the bag.

Place the venison in the bag and coat it with the seasoned water-flour mixture. Add the onions and carrots. Seal the bag.

Put it into a pan at least 2 inches deep. Do not let any part of the roasting bag hang over the edge of the pan. Cut 6 ½-inch slits in the bag.

Into the oven with the works. Bake for 3 hours.

Serves multitudes.

Bon appétit!

FOUR

The Woodcutters

{ A Story of Mutual Profit }

Scarlet runner beans wind their tendrils around the past and bring it into the present. In his 1633 *Herball*, the British botanist John Gerard writes of a large plant not differing from other beans in the way that it grows, but "his floures are large, many, and of an elegant scarlet colour: whence it is vulgarly termed by our Flourists, the Scarlet Beane." Nearly two centuries later, Thomas Jefferson grew them

in his vast vegetable gardens at Monticello. Now I am among those who yearly plant *Phaseolus coccineus*, the "scarlet bean," for its beauty, the flavor of the fruit, and the ease with which the beans may be dried and shelled. Sown in my backyard in mid-May, the seeds sprout in a week, and the resulting vines, flush with heart-shaped leaves, climb their bean towers vigorously. As they reach toward the sky, blossoms like miniature red roses hanging in clusters, they connect me to gardens and gardeners past and present. And they introduce me to some bees that I neither expect nor recognize.

My scarlet runners are abuzz with bees in gently calligraphic flight visiting the flowers. I know what they're doing for themselves and for me. On their own behalf, they seek food. On mine, thanks to their pollinating services, there shall be beans for supper and beans for the freezer. They follow their instincts. We both profit. But just who are these bees? The upper side of the thorax is covered with a vest of furry off-white hairs except for a small, round, dark patch like a tiny mandala in the center. Not honeybees, for they're too big, nor are their abdomens striped; not bumblebees, for they're far too small. These creatures seem to pay no attention whatsoever to my presence. Because my curiosity has been once again roused to a raging itch, I go inside and fetch a small glass jar that once held bouillon cubes. Lid in my left hand, jar in my right, I close in on two of them. Aha! The next day, they lie still on the bottom of the jar. Off to the County Extension Agency we go; there, a young woman puts them into a slender vial, snaps on a lid, and sends them down to Virginia Tech for an ID by the entomologists. The reply arrives within a week: *Xylocopa virginica*, carpenter bees. Some 4,000 species of bees are native to this country, and I get the carpenters.

Carpenter bees as pollinators? I shake my head. Remembering the holes, big as a dime, which they drilled in the cedar fascia of my onetime home in Connecticut, I've always thought of them as destructive pests. We called the exterminator regularly to shoot a liquid toxin into their holes until our county extension agent (hurray for these people!) told us that we could do the job ourselves by putting Sevin in a spray attachment on our hose. But I've discovered no holes in the wooden house in which I now live, and the bees in my bean patch seem benign. Though no other insects have busied themselves around the scarlet runners, I have seen an occasional ruby-throated hummingbird, attracted by their come-hither red, sipping at the flowers, and certainly, a ruby throat's long bill is able to transfer pollen from anther to pistil.

Pollinators—they're so important that the U.S. Postal Service issued a 2007 stamp honoring them, and the U.S. Senate grandly declared June 24–30, 2007, as National Pollinator Week. The stamp shows not just a hummingbird, but a bat, a butterfly, and a bee. Other birds, like orioles, also aid fertilization by the transfer of pollen, as do many other insects, like beetles, flies, wasps, moths, and those pesky little sweat bees that can make a misery of picnics. The wind, too, gives pollen a dizzying airborne transport from flower to flower. Many bees other than honeybees serve as industrious pollinators. Noteworthy among them are bumblebees and orchard mason bees. Wearing a thick coat of golden hair on their abdomens, bumblebees, members of the Tribe Bombini or "buzzers," forage busily for pollen. They are social creatures, building their colonial nests annually in old mouse nests or between walls. They do produce honey but in quantities so small that they are not kept for this purpose. Nonetheless, their buzz has famously inspired Nikolai Rimsky-Korsakov's musical interlude "The Flight of the Bumblebee." Unlike their noisy cousins, Orchard mason bees, *Osmia lignaria*, "scented wood-worker," are modest and solitary, laying their eggs one to a hole, which they cover with a masonry of mud. They, too, have their admirers, although what possessed

Linnaeus to call them "scented" will never be known. Notable among their fans is the French naturalist Jean-Henri Fabre (1823–1915). Giving his young pupils outdoor lessons in surveying and happy to be away from the school's dismal basement classroom (he himself was only eighteen at the time), he noticed that the boys were taking honey from these bees. In his book *The Mason-Bees*, he writes:

> . . . the magnificent Bee herself, with her dark-violet wings and black velvet raiment, her rustic edifices on the sun-blistered pebbles amid the thyme, her honey, providing a diversion from the severities of the compass and the square, all made a great impression on my mind; and I wanted to know more than I had learnt from the schoolboys, which was just how to rob the cells of their honey with a straw.

But the bees make only the honey that each larva requires on hatching from its egg. All unaware, Fabre and his pupils were wiping out a future generation. The orchard mason bees wake early in the spring and proceed to exercise their grand talent, which is the pollination of cherry, pear, apple, and other fruit trees, which open their blossoms before other trees have begun to unfurl their leaves.

Honeybees, *Apis mellifera*, the "honey-making bee," have their aficionados, too, legions of them. Some people lust for the honey; others see their colonies as models of enviably successful communal effort. Prehistoric petroglyphs in Spanish caves show bees swarming angrily around people who are stealing their honey, as do drawings found in India and Australia. The Roman poet Virgil devoted a whole book of the *Georgics*, his long poem on farming, to the "sky's celestial gift of honey" and the bees that make it. Speaking fondly of the colony as a tiny state with great-hearted leaders, he gives detailed instructions on beekeeping. Hives must be located in flower-sweet, stream-watered places. Swarms are made docile tossing on them a handful of dust. Virgil makes much of the kings that lead the bees. Likely, he knew that the life of a hive revolves around a queen, but it would have been impolitic to laud a female creature in a poem that celebrates the world-dominating energies of Augustus Caesar. Virgil also gives a new spin to the story of Orpheus and Eurydice. In his version, Aristaeus, a farmer and beekeeper, attempts to rape Eurydice, who flees but steps on a snake, lurking in the high grass, that inflicts a fatal bite. Aristaeus suffers divine punishment, of course, for his lust and crudeness: his bees sicken and die. But after much travail, Aristaeus makes the proper apologies to the powers of heaven and gets his bees back—not those that he lost but a new swarm, magically engendered in the rotting innards of four sacrificed bulls.

And bees, in Virgil's view, may be able to make the divine manifest:

> . . . some say that bees own a share of the divine soul and drink in the ether of space; for, god invests everything—earth and the tracts of the sea and deepest heaven; from him, flocks, herds, men, all species of wild animals—each one gains for itself at birth its little life; doubtless, afterward, the bees return to him and, released, are made new; death has no place but, alive, they fly up, each to be counted as a star and ascend into heaven above.

In other words, bees may well be immortal, sweetening the heavens with their light as once they sweetened earthly provender.

Two millennia later, in 1901, Maurice Maeterlinck (1862–1949), 1911 Nobel laureate in literature, published *The Life of the Bee*, a loving investigation into the "mysteries of the palace of honey" and its hard-working inhabitants. He even manages to wax poetic on the honeybee's notorious sting:

> There is the distressful recollection of her sting, which produces a pain so characteristic that one knows not wherewith to

> compare it; a kind of destroying dryness, a flame of the desert rushing over the wounded limb, as though these daughters of the sun had distilled a dazzling poison from their father's angry rays, in order more effectively to defend the treasure that they gather from his beneficent hours.

Adulation hardly stops with Maeterlinck. Even though the honeybee is not native to the New World but came here with the early colonists, seventeen states have adopted it as their official insect. And stories about people falling head over heels for them will always pour forth like fresh-flowing honey. Recent years have brought a sweet spate of books. Some of my favorites are *A Book of Bees: And How to Keep Them* by noted science writer Sue Hubbell; Rosanne Daryl Thomas's memoir *Beeing*; and Holley Bishop's wondrously informative *Robbing the Bees: A Biography of Honey, the Sweet Liquid Gold That Seduced the World.* It tells of this creature's biology. It covers private and commercial beekeeping, the making of beeswax, the harvesting of honey, and mouthwatering recipes for dishes like a honey smoothie and a honey burger with onions and jalapeños, it describes the barbed mechanism of the sting and just how bee venom can prod human flesh into a painful allergic response. Nor does it scant the role of honeybees in pollination, a matter that was not truly understood until Charles Darwin's day. Honeybees are often regarded as the supreme pollinators. Commercial beekeepers transport their hives hither and yon in an enterprise called "managed pollination," in which farmers lease the hives so that their crops will set fruit. Beekeepers also truck the hives from one source of nectar to another—from, say, home meadows to a tupelo woods in bloom—so that honey of a distinctive flavor and consistency may be obtained. (One honey, that made from the nectar of rhododendron and mountain laurel flowers, is toxic to people; eating it can lead to flu-like dizziness, heavy sweating, nausea, and whoopsing. Commercial beekeepers don't purvey such stuff. Contaminated honey is either imported or harvested by rank amateurs. But consuming it leads just to misery, not death.)

Honey is produced from any available nectar, which the bees then treat with an enzyme that renders it into a liquid that fungi cannot digest and spoil. The bees then fan the honey with their wings until it becomes so thick that bacteria stick to it and die because they cannot escape their own wastes. Finally, it is capped off so that it does not crystallize. Should nectar as a necessary food be in short supply, the bees can help themselves to a bit of honey in order to survive.

But honeybees and bumblebees are troubled breeds these days. Their numbers decline for reasons not fully understood. Honeybees are attacked by parasitic mites that arrived without a green card in the U.S. as stowaways on imported bees. Bumblebees have also undergone a decrease in numbers, partly because of parasites imported along with European bumbles that were brought here for greenhouse pollination. Both species suffer from viral diseases with horridly descriptive names, like chalkbrood and foulbrood.

A fierce malady affecting honeybees was widely reported in the last few months of 2006: colony collapse disorder (CCD). Ailments with similar symptoms were noticed in the past and variously called May disease or fall dwindle disease, but the appearance of symptoms is not restricted to a particular season. Beekeepers find that grown bees are completely absent from their hives or that there are not enough workers left to care for the larvae, and decimation or total collapse of the colony occurs in a matter of days. Nor is the disorder isolated but has been found in twenty-two states, including Pennsylvania, Texas, and California, and in Canada as well. Is a virus responsible? A parasite? An extended exposure to pesticides used on the crops that the honeybees pollinate? No one yet knows what causes such disasters. And disasters they are, for many commercial crops from apples and peaches to soybeans, cucumbers, and pumpkins depend on pollination by honeybees, not wild bees but bees from hives that have been transported by their keepers. The recent die-off has no precedent. Without the instinct-driven services of honeybees, growers will suffer another sort of CCD—crop collapse disorder. An acquaintance who keeps bees to provide his family with honey says that CCD has not been a problem with hobbyists; he believes that honeybees kept for pollination purposes are subjected not only to the stresses of transportation but to the fumes of oil and burning gasoline. But the apiculturalist at Virginia Tech tells me that hobbyists, too, are losing colonies.

Habitat loss has also played a role in the plummeting of populations. And climate change affects plants and pollinators as the plants shift their ranges and distribution. Bees are hardly the only creatures affected. *Status of Pollinators in North America*, a report issued in 2006, mentions nectar-sipping bats, some butterfly species, and hummingbirds. Studies show that hummingbirds in search of nectar have altered their patterns of migration to follow plants that have moved beyond their usual venues.

The report, however, gives short shrift to that wild, solitary creature the carpenter bee. It delivers the opinion that, because these bees may cause economic damage, they are not likely ever to be listed as endangered species, even if their numbers show precipitous declines. Poor carpenter bees. No one—no one at all—sings paeans to them. No one studies the diseases that may afflict them. Rather attention is given to the damage they inflict on us. No one praises their raiment or thinks that they ascend as stars into the heavens. Maeterlinck expresses his disdain by giving the genus *Xylocopa* only one short paragraph in *The Life of the Bee.* Note the demeaning verb in the first sentence of that paragraph: "The Xylocopae are powerful bees that worm their nest in dry wood." Until I saw them humming busily around my beans and homing in on one red flower after another, they had never suggested to me their role as facilitators of vegetable sex. But of course! They need pollen to feed themselves and their larvae, and how else will they get it other than visiting the floral factories where pollen originates? The arrangement is the Plus-Plus form of symbiosis known as mutualism: each partner performs a necessary service for the other, to the benefit of both. The generations of carpenter bees and scarlet runner beans are assured a future. Nor are other plants excluded from the bees' diligence. The bees aren't choosy. Science calls them "polylectic," meaning that they collect pollen from a wide array of unrelated plants. They just know a good source of sustenance when they see it. Since I first noticed them at work in the beans, I've seen them visiting the panicles of pale flowers on my red currant bushes, the star-like white flowers on my jalapeño peppers, the flowers hanging like lavender bells on the eggplants, and the open-faced golden sunflowers that were planted by birds in the butternut squash patch.

Carpenter bees come in various guises, and, unlike more-the-merrier honeybees, most are solitary. The genus *Xylocopa*, from Greek words meaning "woodcutter," comprises some 500 species and subspecies worldwide, with four native to the U.S. Two are native to the eastern part of the country—*X. micans*, the "glittering woodcutter," and *X. virginica*, the "Virginian woodcutter." It is the latter that comes to caress the scarlet flowers on the beans. A male bee may be distinguished from a female bee by the look of their heads: her head is all black while his sports white markings. And it is she who works at cutting wood. His job is to find a mate in the short springtime of his adult life and inseminate her. Afterwards he hovers in the vicinity of her nest, buzzing aggressively, chasing away other insects that invade this territory, and sometimes dive-bombing a person walking quickly by. I've been strafed while sitting on the back deck of the house that they infested. No need, however, to be afraid of him, for he has no stinger. No wonder that I could catch two males so easily with my bouillon jar! She does have a stinger but employs it only when something really riles her, like being inadvertently touched by a human hand. With a maxim generated by his mason bees but equally applicable to the woodcutters, Fabre puts it nicely: "If you do not tease the insect, the thought of hurting you will never occur to it."

The reason for the female's carpentry is to make a nest. But how does she accomplish such a feat? With her clever mandibles that act as files rasping away the wood when she vibrates her body. She can start from scratch or begin with an existing cavity like a nail-hole. More often, she takes the laborsaving path of simply cleaning out an old nest. She does not eat the excavated wood but, like the tidy housekeeper that she is, brushes it out. One of the signs that wood is inhabited by carpenter bees is the presence of a pile of sawdust on a porch or on the ground. Another clue to her presence is the stain of fecal droppings on the wood below the entrance hole. She doesn't always choose to excavate her nest within a house but will settle happily for a tree, especially if it's softwood, like cedar, pine, or fir. Hardwoods will do, particularly if they're well weathered. Other wooden objects, from fence posts to redwood lawn furniture, have also been subject to her careful drilling. And when the nest is thoroughly cleared out, she constructs brood cells in the hidden galleries. Before depositing an egg in each cell, she stocks it with beebread, a combination of pollen and nectar, that will serve as food for a newly hatched larva. Egg in place, she walls off each cell with a plug of chewed sawdust. After she has made six to ten brood cells in a row, her work is done, and she soon dies. In its shelter, each larva dines well, pupates, and completes its metamorphosis into an adult male or female. After wintering in this cozy hibernaculum, each emerges as a free-flying bee in the following spring. Miracles all around: once again, the bees complete their cycle; once again, my eggplants and scarlet runners are pollinated.

Carpenter bees usually wreak minimal damage on the wood that they worm. But use of old nests year after year means that new galleries are being created annually—perfectly acceptable in a tree but quite the contrary in a house. Galleries have been measured at ten feet in length. And, because tunnels branching off of tunnels are open avenues for bacterial decay, they may indeed weaken the structure that contains them. To control the bees' cunning craftsmanship, we have an array of methods, from the preventative to the lethal. To keep the carpenters away, paint—not stain or wood preservative, but paint—is a true deterrent as long as it's fresh, not worn. Then, where old nest holes exist, they may be sealed with caulk or a glued-in wooden dowel. The bees inside do not try to worm their way out, and that generation will perish. Fabre explains this refusal to breach a second barrier to light and air with his observations of the behavior of his beloved mason bees, which are as solitary in their habits as carpenter bees. Fabre used paper to wrap the single-celled nests of mason bees first with a layer in direct contact with the mud masonry enclosing the cells, then in a cone that left space between the cells and the paper. He writes:

> When the hour for emergence [from a cell] arrives, a stimulus is aroused and the insect sets to work to bore a passage. It little cares in this case whether the material to be pierced be the natural mortar, sorghum-pith, or paper; the lid that holds it imprisoned does not resist for long. Nor does it even care if the obstacle be increased in thickness and a paper wall be added outside the wall of clay: the two barriers, with no interval between, form but one to the Bee, who passes through them because the act of getting out is still one act and one only.

One act and one only—just as the mason bee is programmed by instinct to make a single attempt to release itself, so is the carpenter bee. We could bar its entry into the world not with caulk or a glued-in dowel but simply with paper, if paper were impervious to rain.

Our other defenses against the woodcutters are chemical, like the Sevin that my husband and I learned to spray on our cedar house or the various insecticidal dusts that may be injected into the galleries (while we are as protectively suited as if we were going for a walk on the moon). My own preference would be for sealing off the galleries rather than suffering possible exposure to toxins.

As it happens, I need to take no measures whatsoever. My house, painted light gray with white porch pillars and railings and roasted-pepper-red shutters, shows no signs of excavation—no dime-sized holes, no sawdust, no fecal stains. Live and let live. Though they cannot know it, carpenter bees are safe in my yard.

FIVE

The Inside Story

{ A Tale of Intestinal Fortitude }

E. coli, *Salmonella*, *Campylobacter*, and a slew of equally sickening bacteria make the news these days on a frighteningly regular basis. They and other agents of enteritis are the new tigers under the bed. We don't know when they'll emerge from hiding and strike, raking our innards with their claw-like toxins. Meanwhile, some of us shun the bags of fresh baby spinach in the grocery store, look with a leery eye at the iceberg lettuce salads at the fast-food joint, and wonder about the safety of the potato salads at the family-reunion picnic. Most of us just buy and eat. But sometimes we are subject not only to a decidedly unpleasant malady but also to a national outbreak of scary publicity when sufficient numbers of us are felled by bacterial onslaughts, and some of us die.

What's the inside story on these germs that take up residence in our bellies and guts, some of them temporarily and others fulltime? I investigate on the principle that it's a not a bad idea to know one's enemy. With this principle in mind, I conducted a survey several years ago of the weeds in my gardens. It seems only fitting to become better acquainted with the members of our intestinal flora, given the tidal wave of publicity that they've received. Like weeds, they thrive where they are not wanted.

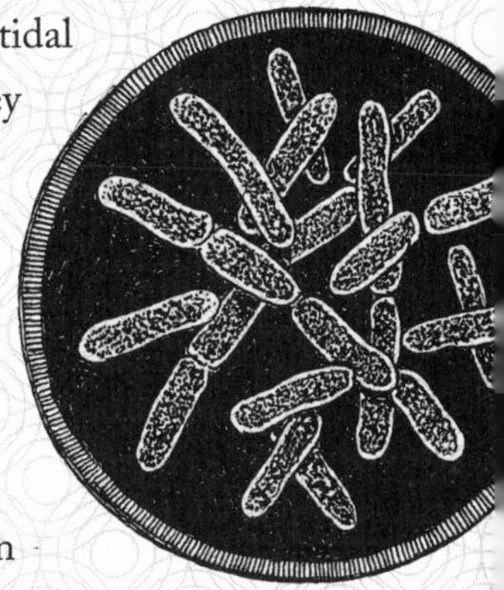

Bacterium—the word comes from classical Greek and denotes a rod, staff, or cane, something large enough, unlike its microscopic namesake, to be seen and held. And two of the bacteria named above, *E. coli* and *Salmonella*, are plump little rods; brightly stained yellow or vermilion so that they can be seen under an electron microscope, they look like slightly elongated jellybeans. *Campylobacter* is skinnier, with a shape resembling a series of wave-like, horizontal S's—a bacterial Gummi Worm. All three are classified as members of the domain named for them—Bacteria, and their kingdom within this domain is Eubacteria, which means "good bacteria" or "abundant bacteria." As we know full well, some of them are far from good; they are, in fact, outright villains. It is as if calling them "good" is supposed to work a reverse magic, propitiating them so that they remain docile and benign. The denizens of this kingdom are characterized by a lack of cell nuclei.

The *Salmonella* species are named for Dr. Daniel Elmer Salmon (1850–1914), who graduated from Cornell University in 1872 and in 1876 earned the first doctorate in veterinary medicine ever awarded in the United States. As the founding director of the U.S. Bureau of Animal Industry, he specialized in studying bacterial disease in animals. One species of salmonella, *S. typhi*, "smoldering salmonella," is responsible for typhoid fever. The food-contaminating bacteria that bear Dr. Salmon's name are found mainly in poultry and pigs. They do not inhabit the innards of a healthy person, but someone suffering from food poisoning caused by salmonella experiences the whole range of symptoms from cramps to diarrhea; it may later lead to arthritis. The most likely victims are babies, invalids, and the elderly. And it takes only a single cell to make someone sick.

How is it transmitted from host to human beings? The Federal Drug Administration gives a partial list of sources in its well-named "Bad Bug Book": "water, soil, insects, factory surfaces, kitchen surfaces, animal feces, raw meats, raw poultry, and raw seafoods." Note that word *raw*. Add peanut butter to the list. In 2007, it was finally identified as the culprit in an outbreak that felled more than three hundred people in 2006. Whether contamination rested in the peanuts or occurred during processing remains a mystery. Eggs, too, are vectors, nor is it just the outside of the shell that may be contaminated but also the inside. An infected hen can transmit the bacteria directly to the yolk before the shell is formed. Forget about making eggnog from scratch; buy a prepared mix. An important secret to avoiding salmonellosis is to give raw foods, from eggs to shrimp, a thorough cooking.

Campylobacter jejuni causes far more bacterial diarrhea than does salmonella. Its binomial translates as "crooked rod of the jejunum," a description that depicts the wave-like shape of its body and its fondness for settling into the jejunum, which is the central division of the small intestine. Like salmonella, it often takes the raw-food route, and it hits hardest at children under five and young adults. Chickens, for one, are likely to carry campylobacter in their intestinal tracts, although the chickens do not fall sick. An estimated eighty to one hundred percent of grocery-store chicken is so contaminated, but, like salmonella, campylobacter's effects, which feel akin to an eviscerating storm, may be eliminated by cooking a bird until it's good and done. Campylobacter may also be found in unpasteurized milk and drinking water that has not been chlorinated. It skulks, as well, in creeks, ponds, and puddles and has been known to inhabit perfectly healthy cattle, birds other than chickens, and flies. Like salmonella, it does not make itself at home in the guts of a healthy person.

But *E. coli* does. These bacteria are part of our normal intestinal flora. They live inside all animals, including cows, horses, pigs, sheep, cats, dogs, and our close kin, the apes and monkeys. And they are always present in our feces, whether we ambulate upon two legs or four. For the most part, they are not only placid but also useful because they ward off damage by harmful bacteria and play an active role in converting the vitamins in our food into forms that we can absorb. The relationship between animal and bacteria represents the kind of symbiosis called protocooperation, in which both sides benefit but in a passive way. The benefits, though real, are unintended and accidental. Neither form of life depends upon the services of the other for survival.

The binomial *Escherichia coli*, "Escherich's creature of the colon," honors Theodor Escherich (1857–1911), a German pediatrician who suspected that bacteria might be responsible for many of his young patients' illnesses and then proceeded to demonstrate that fact. A photograph shows him to be the very model of a Victorian gentleman, stiffly posed, dressed in high starched collar and a heavy, buttoned-up woolen jacket. His hair is parted neatly in the middle and arches on either side of the part, mimicking the curve of his eyebrows; his eyes seem to twinkle behind round eyeglasses; and his whiskers, oh, his whiskers! He sports a long goatee and a silvery handlebar mustache almost big enough to be mounted on a bicycle. Dr. Escherich also published an article on the ways in which intestinal bacteria aid digestion in babies. His discovery of the eternal presence of *E. coli* in human feces led to recognizing that a clean water supply—one not polluted by raw sewage—is necessary to suppress disease, from mild enteritis to cholera and typhoid. Coliform counts monitor the number of bacteria in a water source so that it can be properly treated.

So far, *E. coli* seems to define the "good" in Eubacteria. But it comes in several classes, four of which are raging monsters, the bacterial equivalents of Mr. Hyde. All four classes afflict the innards acutely but each works a variation on that theme. These virulent *E. coli* are enteroinvasive (EIEC, for short), enteropathogenic (EPEC), enterotoxigenic (ETEC), and enterohemorrhagic (EHEC). Those tongue-twisting names translate respectively as "invading," "producing disease," "producing poison," and "causing heavy bleeding," all in the gut. Infection by the first of these, EIEC, leads to dysentery brought about when it invades intestinal tissues that are not furnished with blood vessels. No one knows just what foods may carry EIEC, but some blame has been assigned to the usual suspects—hamburger and unpasteurized milk. Any food tainted by fecal matter from a sick person can transmit the disease. Nor is EIEC picky; it can affect anyone of any age. But its effects are mild compared to those of its gut-wrenching siblings.

The common name of the sickness caused by EPEC is infantile diarrhea, a malady that infects adults, with the exception of travelers, only once in a blue moon. It strikes babies especially hard in emerging countries that lack clean water and proper sanitation. And it works its damage when the bacteria fasten themselves like unbudgeable little stick-tights to intestinal tissues and destroy them. ETEC is responsible for Montezuma's revenge, the diarrhea that flattens people visiting developing countries. Like EPEC, ETEC is not associated with particular foods, except when those foods may have been contaminated by a food handler infected by contact with fecal material. Again, clean water and good sanitation will keep ETEC at bay.

It's the enterohemorrhagic class that has created recent wariness and consternation, not to mention fear. These rod-shaped bacteria, which resemble microscopic vitamin or analgesic caplets, are those designated as O157:H7. It doesn't help to know that this variety is rare, not since more than two hundred people who had consumed raw spinach from grocery-store packages were sickened in 2006, and three of them died. Only a few months later, people eating salads at some Taco Bell restaurants were felled by O157:H7 that probably infested prepackaged iceberg lettuce. Nor is it just fresh greens that carry such bacteria. Undercooked hamburger, raw milk, apple juice and apple cider, alfalfa sprouts, and potato salad have been identified as vectors. Although outbreaks of food-borne enteritis have increased lately, most cases are isolated; they are reported to health authorities—and make a big splash in the media—only when several people become ill from eating the same food from the same source at the same time. More shortly about the source, which is emphatically not the grocery store nor a fast-food restaurant like Taco Bell.

O157:H7 works its inside job by secreting a toxin that ravages the blood vessels in the intestines, kidneys, and brain. The symptoms are a veritable intestinal tsunami—cramping and diarrhea, watery at first, that becomes very bloody. Often, there's no fever. Three to eight percent of O157:H7's victims suffer kidney failure and a gross savaging of the red blood cells. Robert Tauxe of the Centers for Disease Control gives a graphic description: "The *E. coli* toxin damages blood vessels by creating small strands across the insides, so when the red blood cells go through them, it's like they're going through a cheese cutter. It just slices up the red cells." He adds that taking antibiotics or an antidiarrheal product like Imodium can make the enteritis worse because the germs, opting for life as do all living things, may counterattack by secreting even more toxin. Though enduring enteritis may well make us feel as if death would be preferable, we usually recover, thank goodness, without needing medication.

We can take measures to minimize the possibility that O157:H7 will lay us low. All are commonsensical. Drink pasteurized liquids. Cook meat thoroughly; an internal temperature of 160 degrees will kill *E. coli.*

Store fresh vegetables and fruits in a fridge set at forty degrees or lower. Buy cut produce, like melons or salad greens, that's been refrigerated or iced down; frozen or canned foods are all right, for they have been processed in ways that eliminate any contamination by *E. coli*. Put leftovers into the fridge no more than two hours after cooking and throw them out after four days. Don't eat any store-bought alfalfa sprouts, but the home-grown variety should be safe. Use one cutting board for meat, another for vegetables. And wash hands thoroughly with soap and warm water for twenty seconds before and after preparing food. (In my experience, it takes a good twenty seconds or more to rinse off a liquid soap completely.) As for washing fresh greens, that's a good idea, for it will wash away residual pesticides. But it won't remove O157:H7. That superminuscule critter sticks to greens more tightly than bubble gum does to the sole of a shoe. Once it's there, it's there for good—or, to put it more accurately, it's there for bad. We are aware of gum on a shoe, but because bacteria are invisible, they are doubly insidious.

These measures are precautions, not panaceas. They do not guarantee that we won't pick up a bag of contaminated greens at the supermarket or buy an undercooked hamburger at a restaurant. They are meant only to help maintain sanitary conditions in our home kitchens. But we're not sitting ducks waiting to be knocked down by the next outbreak of bacterial enteritis. To begin with, the incidence of food-borne sickness caused by ground beef, milk, and juice has decreased dramatically in the Western world because of irradiation and pasteurization.

Irradiation—the very word evokes shudders in many people. And a powerful mythology has risen around it. The Centers for Disease Control says, however, that irradiated food does not become radioactive, that dangerous substances are not produced by the process, that the nutritional value of the food is not changed, and, most important, that disease-causing germs are reduced or eliminated. Those germs include not just *E. coli* but also *Salmonella*, *Campylobacter*, and a mess of others. The process also does in many parasites and molds. Raw meat and poultry are the foods most frequently treated; for these, both the Federal Drug Administration and the U.S. Department of Agriculture must grant approval. Other foods, some of them surprising, also undergo irradiation: wheat flour to prevent mold; white potatoes to inhibit sprouting, herbs and spices to sterilize them. The process has been applied, as well, to Florida strawberries to prolong shelf life and Hawaiian fruits, formerly fumigated, to eliminate fruit flies that could cause harm on the mainland. And all this has been going on for more than forty years with wheat flour, approved in 1963, first out of the starting gate. The process itself was discovered in the 1920s by French scientists. NASA makes sure that astronauts eat food sterilized by irradiation. Nor does irradiation stop with things edible. Anyone with a knee or hip replacement wears an irradiated joint. How does irradiation proceed? By one of three methods: exposure to the gamma rays of radioactive cobalt or cesium, electron beams, or X-rays. The last two methods involve no radioactivity whatsoever. Irradiated foods sold in grocery stores must be labeled "treated with radiation" or "treated by irradiation" and bear, as well, an international symbol, the Radura. The Radura looks like a green life ring surrounding a flower. The name, an invented word like Kleenex or Lysol, is based on the Latin *radiare*, to gleam, emit rays, radiate. Nonetheless, a glance at the strawberries, beef, and poultry at my supermarket, one of a large chain, discovers plenty of USDA Inspection labels but not a single Radura.

If irradiation can keep astronauts from getting sick, give strawberries a longer shelf life, and eliminate disease-causing microorganisms from chicken and beef—letting us eat processed hamburger medium rare!—it could also its work magic on fresh greens. The USDA labs in Pennsylvania are now exploring how irradiation affects the greens we love. It does not turn greens to mush, though flavor might be slightly altered, just as pasteurization alters the flavor of milk. (Having drunk raw milk on my father's farm, that's a change that I can easily go along with, for when cows are turned out to pasture in the spring, they will eat onion grass, which gives the milk a most unappetizing flavor.) But there are good reasons for not irradiating greens. One is the widespread, albeit mythic, perception that eating such stuff is the equivalent of putting our bodies directly under a mushroom cloud. And if people think that something is dangerous to health, it doesn't sell. Another reason is that the FDA must approve the process for each specific application—spinach here, sprouts there. Yet another reason is that we do not now have the capacity for irradiating the thousands of tons of leafy greens grown each year. Installing processing equipment available to every vegetable farm and packaging operation would involve spending boggling sums of money to the detriment of who knows what other worthy causes.

What can be done to keep greens free of bacteria? To keep the unseen, uncountable multitudes of bad bugs in check? Go to the source of the problem. Clean up the farms where fresh produce is grown. Contamination often results from irrigating crops with water that contains animal wastes and their bacterial hitchhikers. The source of the spinach-driven *E. coli* outbreaks of 2006 was traced to wild pigs roaming at random in California's Salinas Valley. Months afterward, though the USDA pronounced spinach clean, sales dropped precipitously, and California, which supplies a full seventy-five percent of the leafy vegetables sold in the U.S., took a huge economic hit. As a result, the state instituted a voluntary inspection program early in 2007. The first participants comprised twenty-four companies that wash, package, and ship some seventy percent of the greens grown in the state. The companies that process greens and the shippers have agreed to buy produce only from farms that have protected their crops against bacterial contamination by taking such measures as building fences to keep animals out of the tender rows and testing irrigation water on a regular basis.

A caution: the label "organic" does not ensure safety any more than would pasting a smiley face on a package. In the case of leafy greens, "organic" means that no chemical fertilizers, herbicides, and pesticides have been applied to the soil of the farm or to the produce as it grows. It does not mean that irrigation water or the handlers are free of germs.

But the inside story can have a happy ending. There are two ways to meet the challenge of finding greens that will not cause one form or another of Montezuma's revenge. One is to trust the USDA's assessments of risk: eat what's approved, and skip the rest. The other, the one that speaks loud and clear to my sensibilities, is to stay local. We can patronize our farmers' markets, grow vegetables in our own gardens, or both.

Here's to good health.

SIX

Rascal

{*Cases of Entanglement*}

Walt Whitman has famously said, "A morning-glory at my window satisfies me more than the metaphysics of books." For the most part, I disagree—and suspect that Whitman has never lived with morning glories on the rampage. For me, a morning glory at the window signals hard labor, and I'd far rather read a book than deal with those obstreperous vines.

In my yard, the morning glory flowers are glorious indeed, opening their trumpets—white, deep purple-blue, and pink, big and small—to the morning sun. But the plants define entanglement. They climb the pole bean towers and the rose of Sharon bushes. They wrap themselves around chain-link and tomato cages, around the tomato vines themselves, and around the chicken-wire fences that keep groundhogs away from Kentucky Wonder seedlings and the fruit of butternut squash. Self-sowing, they are well nigh impossible to eradicate. Because they are annuals, they die off, and each spring, I find myself disentangling the dead vines from all that they have beribboned. The flowers are beautiful; the plants are pests.

As long as I seem doomed, year after year, to give morning glories room to entwine themselves lustily around, around, and around every possible vinehold, I feel bound to take their measure. The relationship of plant and woman often seems a Plus-Minus, with morning glories winning and woman losing every time. But when I take my hoe in hand, I assume the Plus role of a predator, woman versus weed. And I can keep my prey, the vines, in check. Yet, when I strike, the well-being of this overexuberant plant is enhanced: each piece of severed root is capable of sprouting a whole new vine all on its own. The morning glory is truly a Hydra among plants, like the mythical monster that grew two heads for every one

cut off. The hero Hercules eventually killed it by sealing the stump of each severed head with its venomous blood. I could replicate this feat by using a modern venom in the form of a herbicide, but it won't do to poison the garden's other denizens. Best to let the Hydra have its way.

Once upon a time, though, a particular morning glory surprised me. Magically, almost overnight, the tomato cages in my North Carolina garden were covered with slender vines bedecked with delicately ferny green leaves and tiny flowers shaped like scarlet stars. I took a photograph and sent it off to a botanist at the North Carolina Maritime Museum. The answer came back overnight: *Ipomoea quamoclit*, which is less formally called cypress vine or hummingbird vine. The species name is supposedly the Mexican word denoting the vine. The first common name refers to the leaves' resemblance to the soft, flat, flexible, feather-like needles of bald cypress; the second, to the hummingbird-beckoning red of the flowers. In summer, the birding and botanizing visitors who visited my garden invariably took home packets of black seeds resembling small peppercorns.

Linnaeus himself (1707–1778) named this genus. It comes from two classical Greek words, *ips*, which means "worm," and *homoios*, "resembling." Worm-like—what an apt simile for those lively, almost wriggling vines! And the genus is huge, comprising more than five hundred species, including some shrubs and trees. Bush morning glory (*I. leptophylla*, "thin-leaved"), a native perennial shrub with plump tuberous roots, grows in a broad band from southern Montana to northern Texas. Morning glory tree is the common name of *I. arborescens* ("woody," "tree-like"), which is native to Mexico; up to thirty feet in height with a crown spread of twenty-four; it sheds and re-grows its leaves according to the season. The family to which these vines, shrubs, and trees belong is the Convolvulaceae, the "wind-arounders" or "entwiners," which includes the morning glory-like bindweeds, *Convolvulus* species, that give the family their name. Some of the *Ipomoea* are native to North America, while others were introduced from Central and South America. Cypress vine came north from Mexico. Some are annual; others, perennial.

Every state except for Idaho and Alaska is thoroughly entangled. Arkansas has declared the morning glory to be a "noxious weed," while Arizona has gone a giant step farther, pronouncing it a "prohibited noxious weed." By way of enforcement, seed companies are not allowed to ship to Arizona customers the seeds of any variety. For them, no heirloom 'Heavenly Blue' with its pale yellow throat and golden anthers, no dark crimson 'Scarlett O'Hara,' no magenta 'Mini Bar Rose' with a silvering of white on flower's edge. I doubt that Arizona will ever eliminate this scourge, for a wild menace has no cognizance of civilized entreaty but rather rears itself up wherever its seeds happen to alight and wherever its roots are cut, making two plants of one. Aside from which, Arizona has its share of native species, including the canyon morning glory (*I. barbatisepala*, "hairy-sepaled"), the purple morning glory (*I. capillacea*, "resembling a hair"), and the triple-leafed morning glory (*I. ternifolia*, "three-leaved"), and other native species lurk in California and New Mexico, just over the state's most pervious borders. And some *Ipomoea* species seem not to be *verboten* in the Grand Canyon State, among them the moonflower (*I. alba*, "white") and the morning glory tree. The latter is known in Spanish as *palo del muerto,* "tree of death," but its gnarly branches and the great white flowers with which it covers itself bespeak life. Most likely, Arizona exempts these two because they are not able to escape captivity but must be coddled.

Moonflowers are the contrarians of their genus. They might be called evening glories, for, unlike their cohorts that begin to unfurl their tightly closed buds at the break of day, these denizens of the night open their flowers at twilight. In warm weather, moonflower buds take only a few minutes to explode into fragrant, almost incandescent white blossoms as much as eight inches in diameter. When dawn arrives, the petals wilt and close. In hot climates, the vine is a perennial but may be grown from seed as an annual elsewhere. And, oh, how it grows! Give it a trellis or a tower to climb so that the flowers on its long vine may more closely inspect the stars and its namesake moon.

Admire the flowers—or dislike them, inhale the perfume of the scented varieties, be astonished at the bush and arborescent forms, but know that morning glories can be poisonous. The American Society for the Prevention of Cruelty to Animals maintains a database of plants toxic to pets that lists *Ipomoea* species in general and shows pictures of *I. tricolor* ("three-colored"). That's the species of which the heirloom 'Heavenly Blue' is a cultivar; every one of the seven flower-seed catalogues in my current collection offers it. But most dogs, though driven at times to sample grass right heartily, are not much interested in eating plants, nor, I suppose, are many cats, though there are exceptions. My tiger tomcat engages extensively in eating green veggies—lettuce, broccoli leaves, beans. Houseplants, also prey to his appetites, have been banished from my quarters; so, I wouldn't put it past him to nibble on morning glories. He's not given the chance, however, for he's an indoor animal, and happily so.

It was the flower children of the 1960s who rediscovered something that Indians in Mexico had known since time immemorial: the seeds of morning glories can be made to yield a hallucinogenic substance, LSA or d-lysergic acid amide, a close cousin of LSD. Like LSD, it can take the person who ingests it on a dizzying, sometimes nauseating psychedelic ride. A columnist for the DIY Network on Home & Garden Television says that she experienced these giddy effects accidentally, along with a long-lasting, headachy hangover, through contact with the plant's sap when she ripped out a thickety tangle of vines. With Mexico's Zapotec Indians, the substance was used ceremonially on behalf of the common good. But for many others, barring the poor woman stricken by sap, the aim seems to have been attainment of a personally blooming and transcendent high. I'll rest content with the descriptions of delirium, for I'm not about to put this kind of intoxication to any test.

In addition to recreational and ceremonial use, some species of morning glories have seen service in the vast pharmacopoeia of herbal remedies. Chief among them is *I. purga*, the "purge," which grows in Central and South America. Its deep pink flower would cover my hand. The species name gives away the plant's nature—it's a laxative, and a powerful one, to boot. It has also been employed as a vermifuge to destroy internal parasites like pinworms. The common name of this innards-scouring morning glory is jalap, after the Mexican town of Xalapa, which has also bestowed its name upon the palate-scouring jalapeño pepper. The resin from which the laxative is made comes from jalap's root, which develops round tubers that may be as small as a walnut or as big and plump as a turnip. It does not seem to figure, however, in modern medicine.

Pests! Purveyors of poison! Purgatives! Yet, there's one morning glory that we'd be hard-pressed to do without: *I. batatas*, the sweet potato. It's hard to imagine life without candied sweet potatoes at Thanksgiving and a sweet potato pie at any time of the year. Aside from which, being full of the antioxidant beta-carotene, plus a rich share of vitamins C and E, they are good for us. And they were anciently appreciated. They were probably domesticated some ten thousand years ago in Peru—the plant decorates prehistoric Peruvian pots, and cultivation had spread throughout South and Central America into the Caribbean in pre-Columbian times. In 1492, when the grand explorer stepped ashore on Hispaniola, thinking that he'd reached India, he saw—and likely sampled—the sweet potato. The following year, he returned to Spain with this vegetable treasure in his cargo. His ship took *batatas*, the vernacular name by which the plant was known in Hispaniola, across the sea, where it was given residence in the royal gardens of Ferdinand and Isabella. A short two decades later, Peter Martyr d'Anghiera (1457–1526), reported that sweet potatoes grew in Honduras. Peter Martyr was an Italian who served as a chaplain in the court of Ferdinand and Isabella and was later appointed as the Spanish Royal Chronicler by the teenaged Charles V (1500–1558). In his quest for details of Spanish explorations, he was able to speak directly with Columbus and other bold adventurers. In 1530, four years after his death, his book *De Orbe Novo—On the New World*—was published, and it endures to this day, a splendid record that gives details on matters from food to the nature of the encounters of the invaders with the native peoples. Of course, the Spaniards took home not only sweet potatoes but also seeds and specimens of every exotic plant that they encountered—zinnias, sunflowers, marigolds, corn, beans, squash, and many more—and introduced them to Europe. The Spanish also took sweet potatoes as far as the Philippines, and the Portuguese brought them to Africa.

But even before the Old World marched in clanking armor through the New, sweet potatoes made their way, Kon-Tiki fashion, to Pacific islands like Tahiti, Fiji, the Cook Islands, and the Marquesas, and were put into cultivation by the Maori in New Zealand as early as the 1300s. The sweet-potato names found across the Pacific are kissing-cousin close to their South American counterparts. One garden historian writes: "The Maori name for sweetpotato [sic] is *kumara*, which is very similar to a common South American name. Based on Maori lore, if the sweetpotatoes were growing poorly, farmers would place wooden *kumara* gods and dried human heads in their fields to improve the yield." So, as well as providing nourishment, sweet potatoes have, at least once in history, exacted a human toll.

In Europe, sweet potatoes were also costly but in a way lethal only to the pocketbook. They commanded high prices in the early 1600s at England's markets because of their supposed aphrodisiac powers. Though he was unacquainted with ornamental morning glories, the English botanist John Gerard (1545–1612) writes admiringly in his *Herball* of the sweet potato, to which he gives the scientific names *Batata Hispanorum* and *Sisarum Peruvianum*, "Spanish Potato" and "Peruvian Skirret." (A skirret is an Asian herb with an edible tuberous root.) He distinguishes this vegetable from the white potato, which he calls *Batata Virginiana*, the "Potato of Virginia," after the place from which he got his seed potatoes. Unknown to Gerard, these so-called Virginia potatoes actually originated in the Andes and were relished by the Incas. As usual, the Spaniards came, saw, ate, and took them home. What the Incas called them I do not know, but because they looked something like the batata, that name slid, in the 1600s, from its rightful owner onto a rounder and lumpier tuber. But, while sweet potatoes were readily accepted by Europeans from the moment of their introduction, white potatoes weren't widely used as a foodstuff for two more centuries. The reason: their leaves identified them as members of the Solanaceae, the Nightshade family, which comprises many plants containing solanine, a poisonous alkaloid.

Gerard commends the Spanish Potato's virtues with lavish praise (and evidences belief that Columbus did indeed reach the nation called India). "The Potato roots are among the Spaniards, Italians, Indians, and many other nations common and ordinarie meate, and do strengthen and comfort nature." He recommends roasting them, then sopping them in wine or seasoning them with oil, vinegar, and salt. And he gives a ringing endorsement to their aphrodisiac powers: "Howsoever they be dressed, they comfort, nourish, and strengthen the body, procuring bodily lust, and that with greedinesse." The Spanish Potato, by that name, entered what is now the United States from Cuba or Jamaica and began to be grown here as early as the mid-1600s.

Gerard does not mention the flower nor show it in the woodcut that accompanies his description. Sweet potato blossoms are rarely seen, for the plants are grown from stem cuttings, but when they do appear, they have the trumpet-shaped flower characteristic of morning glories. It gleams white with a deep pink throat. Ornamental sweet potato vines often appear in landscape designs. Their flowers are brighter and their leaves more colorful than those of the food plant. The cultivar 'Marguerite' spills big chartreuse leaves down the retaining wall between the sidewalk and the land on which my local library stands. The leaves of another cultivar, 'Blackie,' are midnight purple. These ornamentals develop tubers, but woe be unto anyone who takes a bite, for they are the bitter antithesis to the food plant's inborn sweetness.

Almost as soon as the Spaniards reached home shores, morning glories appeared in European flower gardens. Like zinnias and marigolds, they did not move north from Central America but twice crossed the Atlantic—eastward to Europe, then westward to North America. Two species were notable in American gardens. One of them, *I. purpurea*, remains rambunctiously established to this day. Though *purpurea* means "purple" or "crimson," the flower's range of color runs riotously from snowy white to magenta to purple with a five-pointed crimson star radiating out from the center of the petals. *I. purpurea* is known colloquially as the common morning glory, although the USDA, for reasons that I cannot guess, refers to it as the tall morning glory. Its cultivars, many of them heirlooms, bear a wonderment of names, including 'President Tyler,' 'Grandpa Ott,' and 'Cameo Elegance.' 'Grandpa Ott,' a

variety originally grown in Bavaria, was given to Diane Ott Whealy and her husband by her grandfather, Baptist John Ott, who lived on a forty-acre farm in Iowa. It is now proudly offered by the Seed Savers Exchange, which promotes its ability to self-seed. And here's the description of 'Cameo Elegance' given in the Spring 2007 catalog of the Territorial Seed Company:

> The variegation of Cameo Elegance brings allure to the garden. The radiant jade and cream marbled foliage is ornamental enough to stand on its own, but add the dazzling white-throated fuchsia blossoms, and you've got the most picturesque combination.

I shake my head vigorously. Self-seeding! Allure! These are nothing more than sales pitches designed to mislead the unwary catalog enthusiast into buying a pest.

The earliest recorded mention of *I. purpurea* in the American colonies (though not as a pest) was made by John Bartram (1699–1777), who was not only the botanist appointed by George III to keep track of New World specimens but also a friend of Benjamin Franklin. In the 1800s and early 1900s, the plants were beloved by pioneers, whom the flowers satisfied as surely as they did Walt Whitman; old photos show the vines growing up wires or strings in front of log cabins and sod houses. By the early 1900s, however, this morning glory began—justly, I think—to assume the status of a weed. In 1910, Harriet Keeler, amateur botanist, educator, and author of a book on tree identification, issued this no-nonsense dictum in her book *Our Garden Flowers*:

> Somewhat of a rascal when given too free a hand in the garden, it must be kept within bounds or not kept at all. A flower that is seen at its best only at sunrise, it will never be a favorite with the American people, and the morning glory

> vine, despite all its virtues, will probably remain a utility plant loved by a few, outlawed by others, tolerated by the many.

Rascal? That's a weasel word! It vastly understates the situation. The common morning glory has crashed my garden party by winding itself around, around, and around every possible fence wire, pole-bean stem, and shrub in my backyard. It embraces them more tightly than the arms of any lover. And, in its multitudes, it can't be uprooted lest the Hydra effect come into play. My only recourse is to catch the vines as they begin to climb and to tear them down before the heart-shaped leaves bedizen everything. The presence of the common morning glory summons a four-letter word: Work.

The other species favored in the U.S. from the end of the eighteenth century on was first mentioned by none other than Thomas Jefferson. On December 16, 1790, from Philadelphia, he wrote this to his son-in-law, Thomas Mann Randolph, husband of Jefferson's daughter Martha Washington, better known as Patsy:

> I send herewith some seeds which I must trouble you with the care of. They are the seeds of the Sugar maple and the Paccan nuts. . . . There is also in the same tin box some seeds of the Cypress vine for Patsy.

In March of the following year, Patsy responded to her father, still in Philadelphia, that she and her sister Polly had planted the cypress vine seeds in window boxes at Monticello.

Oh my! The delicate vine with ferny leaves and scarlet flowers that so surprised and delighted me in my North Carolina garden is not just a wildflower but an heirloom, a plant to be treasured. And how well it will match another of Thomas Jefferson's favorites that I grow religiously for its colors and its flavors, the scarlet runner bean! Would it also attract the pollinators, the carpenter bees and hummingbirds, as does the bean? There's only one way to find out.

I place an order for a packet of seeds.

SEVEN

A Bale of Chelonians

{*Tales of the Licit and the Illicit*}

The turtle lives 'twixt plated decks
Which practically conceal its sex.
I think it clever of the turtle
In such a fix to be so fertile.

—Ogden Nash

Bale is the collective noun—the vener-able venereal term—that describes a gathering of turtles, just as whales gather in a pod and fish in a school, while a herd of cattle grazes in the field and a clowder of cats most certainly will give birth to a kindle of kittens. A bale is the kind of arrangement in which Dr. Seuss's Yertle found himself at the bottom. But Yertle's case simply exaggerates the natural tendency of freshwater turtles to pile up. And chelonian is the word that covers all of turtlekind. In this story the bale comprises only eight, most of which have not encountered the others in real life, but here I take the liberty of piling them up in a tale of turtles. Before you read their stories, let me give you their names: George, Timothy, Crash, Splash, Carolina, Hambone, Lilypad, and Wumpus—especially Wumpus. As their stories unfold, you shall also meet the Turtle Police.

First, a short digression. Chelonian—the word, which comes from the classical Greek word for "tortoise," applies to the seagoing turtles like the leatherback, hawksbill, and loggerhead, that have flippers. It denotes, as well, the freshwater turtles, like snappers, sliders, and cooters, that use their webbed feet to paddle around in ponds and creeks. Though the box turtle may seem completely terrestrial, lacks webbed feet, and wears a tortoise-like dome, it too is classified with the freshwater group because it needs to live near water for drinking and soaking in, especially when it's a hatchling, even if the water is just a swamp. (The box turtle's binomial *Terrapene carolina*, "the terrapin from Carolina," would seem to indicate that it's a terrapin, but the word *terrapin* comes from an Algonquian Indian term meaning "edible turtle living in water." *Turtle* and *terrapin* are synonyms, despite the *Webster Collegiate Dictionary*'s pronouncement that the latter word denotes an edible species. The definition-maker may have been thinking of the diamond-backed terrapin—*Malaclemys terrapin*, "marsh turtle"—an estuarine, crab- and snail-eating species, which was hunted nearly to extinction in the early 1900s as a prime ingredient for soup. But no

member of turtlekind is inedible, though some may not taste very good.) Chelonian refers to the tortoises, too, including Florida's gopher tortoise, the Burmese mountain tortoise, and the giant tortoise of the Galápagos. All tortoises are digitigrade creatures—that is, they walk on their tiptoes, like cats and dogs. They are thought of as strictly terrestrial animals, but some, like the Galapágos giant, do not spend their entire lives high and dry but enjoy a bit of soaking in the surf or a nearby waterhole. The Roman name for turtle or tortoise is *testudo*, which means "shell," and the word is used to this day for the Testudines, the order to which all turtles and tortoises belong.

George was the first turtle that intersected with my life. Which is not to say that I ever got to know him more than superficially; rather, he constituted a somewhat alien presence in constant need of food and habitat-housekeeping. When we brought him home from the ten-cent store in Whitefish Bay, Wisconsin, he was a hatchling about as big around as a quarter, and he couldn't have been more than a few days old, for the egg-tooth used to break out of his shell was still attached to his beak. We did know that he was a red-eared slider but we didn't know then whether he was really a male turtle—the sexing of a slider needs to wait until the turtle's shell reaches a length of four inches stem to stern. We named him George anyhow, and as it happened, the name fit. He grew a little, he grew a lot, and as he grew, his foreclaws lengthened into long, sharp, slightly curved needles. Female foreclaws are far more stubby. In his earliest days with us, his care and maintenance became my responsibility and no one else's, for whichever child had originally craved a pet turtle had almost immediately lost all enthusiasm—turtles aren't soft and furry, they don't purr or wag their tails. As George aged, he became ill-tempered, or so it seemed to me, his keeper. He would thrust his head forward and hiss when I approached. He was given to attempting escape as well. I now know that he was most likely seeking an accommodation more suitable than a large dishpan fitted out with a rock and a slosh of water.

A rule of thumb, of which I was unaware back then, is that the habitat for an aquatic turtle should have a length of at least five times carapace length, a width three times carapace length, and a depth of two times carapace length plus twelve inches to prevent escapes. George did not have sufficient room for swimming. He kept growing nonetheless. When we moved from Wisconsin to Connecticut, George, contained in a coffee can held on a lap, flew right along with us. And he kept on growing. When he reached eight in years and six in inches of carapace length, I had had enough of changing his feces-fouled water every day, enough of being hissed and lunged at. I carried him down to the Aspetuck, the stream that flowed at the bottom of our backyard, and released him. Wondering if his years in captivity rendered him unfit to survive on his own, I think of him in the past tense, but he could have made it on his own to this day, forty years later. Captive red-eared sliders have been known to live for more than one hundred years but I suspect that the average lifespan in the wild would be somewhat shorter.

Officially, George was a member of the kingdom Animalia, the Phylum Chordata, the Class Reptilia, the Subclass Anapsida, the Order Testudines, the Family Emydidae, the genus *Tachemys*, the species *scripta*, and the subspecies *elegans*. Translated, his lineage turns into this: an animal with a backbone; in particular, a reptile, with no temporal openings in its skull, that belongs to the shell-wearing order, the family of freshwater pond and river turtles, and, most specifically, the neck-turtle genus's inscribed species that is elegant.

Whence comes a turtle like George and all the world's other chelonian species? The ur-turtle, which was not quite yet a true turtle, lumbered out of the Permian seas some 270 million years ago and became a landlubber. It was a pareiasaur, a "cheek-lizard," of considerable size and heft, shaped something like a pudgy torpedo with eyes and stocky legs. Protective bony plates jutted from its back. Why the nomenclators decided to call it a cheek-lizard is something that I cannot guess at. But it's known that this particular reptile had teeth, which it used to chew its greens, for it was an herbivore.

Somewhere, as turtle-like creatures stomped and swam into the Jurassic period, beginning about 187 million years ago, they lost their teeth and developed sharp, horny beaks. Their ribs migrated outward until the animals were enclosed in the bony house that travels with them wherever they may go. Top and bottom, carapace and plastron, a plate of bone lies beneath each keratinous scute (keratin is the material forming human fingernails and hair, the scales of fish and snakes, the horns and hooves of cattle and goats), and chelonian shoulders became anchored within that bony rib cage. During the Jurassic, many of the proto-turtles left land behind for an aquatic way of life. Two distinct groups evolved along the way, the pleurodires, "side-necks," and the cryptodires, "hidden-necks." The former, now found only in the Southern Hemisphere, do not retract their necks as do the cryptodires but rather curl them around sideways in order to tuck their heads under the shell. In these turtles, the carapace is fused to the plastron. They are the oldest kind of true turtles, and from them the family tree branched, producing the cryptodires, of which the some of the earliest members were the chelydrids—the snapping turtles, followed first by the sea turtles and then by the others in their numerous varieties—the sliders, the cooters, the soft-shells, the tortoises, and all the rest.

Cryptodires have a specialized joint that lets the neck double back in an S-shaped curve so that the head can be pulled straight back into the shell. Their plastron is sutured to the carapace by bony structures called bridges. And cryptodires, though more recent than the pleurodires, go considerably back in time. A fossilized specimen dating back to the early Jurassic was found in Arizona in the late 1980s. Its skull shows the jaw mechanism characteristic of the group. Thirteen present-day families of chelonians exist, and of them, eleven consist of the hidden-necked kind.

As for "turtle," nobody knows the origin of the word. It may have been a corrupt version, mispronounced by British sailors, of the French *tortue*, tortoise. Or, as *The Oxford Dictionary of English Etymology* notes, the cleric Samuel Purchas (1575?–1626), editing the reports of explorer Richard Hakluyt (1552–1616), indicates that the word may be a native Bermudan name for the animal. In any event, early naturalists were somewhat bollixed by this animal that wears its own house. A fifteenth-century manuscript in the Bodleian Library states: "The tortuge is acounted amonge snailles for he is closed bitwene twey hard schellis."

One of the hazards facing a wild turtle is that of captivity or worse. Ever since mankind appeared upon the planet, turtles and tortoises have been scooped up and carted away from their accustomed territories. Some have been destined to become dinner; others, pets; and yet others, items for trade. John Lawson, the English surveyor of the Carolinas, focused on the dinner-worthiness of turtles in his 1709 account of the natural wonders, from trees to birds and fish, which are to be found in the land that he called the "Summer-Country" (a real-estate come-on, if there ever was one, to his fellow Britons). Classifying turtles and snakes as insects because they lay eggs, he describes both sea turtles and two kinds of "Terebins," those that are mainly terrestrial and those that paddle merrily in lakes and rivers. He declares that Land-Terebins, relished by the Indians "are good Meat, except the very large ones, and they are good Food too, provided they are not Musky." His assessment of the water sort is worth a full quote:

> Water Terebins are small; containing about as much Meat as a Pullet, and are extraordinary Food; especially in *May* and

> *June*. When they lay, their eggs are very good; but they have so many Enemies that find them out, that the hundredth part never comes to Perfection. The Sun and Sand hatch them, which come out the Bigness of a small Chestnut, and seek their own Living.

Turtles put to gustatory purposes are not given names. But Timothy is another story altogether, an eighteenth-century tortoise that figured both as merchandise and pet. This chelonian gained some small renown as one of the creatures studied by English naturalist Gilbert White (1720–1793), who wrote *The Natural History and Antiquities of Selborne*, published in 1789. And just who was Timothy? A high-domed, chunky-limbed, strictly terrestrial animal, more formally known as *Testudo graeca ibera*, "Greek tortoise from the Caucasus," a species native to Cilicia, an ancient region of Turkey bounded on the north by the Taurus Mountains and on the south by the Mediterranean. And that, indeed, was not only Timothy's natal land but also the place in which the tortoise spent—a reasonable guess—at least the first ten years of life. So, how did the Rev. Mr. White find an opportunity to make direct observations of Timothy's behavior from cold-weather hibernation to choices of food? Timothy was kidnapped in 1740, that's how; transported to England; and sold for half a crown by a sailor in need of pocket money. Though no one knew how to determine a tortoise's gender at the time, this particular tortoise was female, as can be ascertained by her still extant carapace, which is kept at the British Museum of Natural History. After her arrival in England, she spent the first four decades of her exile in the brick-walled clerical garden of the Rev. and Mrs. Snooke. On the long-widowed Mrs. Snooke's death, Mr. White, who was her nephew, dug Timothy out of hibernation. He describes the occasion in a letter, written in 1780:

> The old Sussex tortoise, that I have mentioned to you so often, is become my property. I dug it out of its winter dormitory in March, when it was enough awakened to express its resentments by hissing, and, packing it in a box with earth, carried it eighty miles in post-chaises.

(Hissing—shades of George!) Timothy's new home was a much less confined clerical garden—flowers, herbs, lots of delectable lettuce—in Selborne, Hampshire, White's hometown, where he served as curate at St. Mary's Church. It had become fashionable in those days to keep an exotic tortoise as a pet. I'm reliably told that tortoises may still be found in many English gardens as if they were living gnomes.

Mr. White found his newly acquired animal a source of both fascination and pity. He observed the rhythms of tortoise behavior from spring awakening to voracious summertime eating to the autumnal excavation of a winter resting place. He weighed her annually and fretted if she'd lost an ounce. He dunked her upside down in a tub of water to see if she could swim—she could not. He considered her habit of hibernation and wondered why Providence should waste longevity on a creature that spent "more than two-thirds of its life in a joyless stupor, and to be lost to all sensation for months together in the profoundest of slumbers." Calling Timothy "an abject reptile" and a "poor embarrassed reptile," he saw her as encumbered by heavy armor and imprisoned in her own shell. (A modern riff on Timothy imagines her describing humankind as "great soft tottering beasts. *They* are *out*. Houses never by when they need them.") Timothy is said to have died in the spring of 1794, the year after her keeper's death. Her years had then amounted to at least sixty-four.

*

Longevity and turtlekind seem practically synonymous. The majestic Galápagos tortoises, wearing shells like statehouse domes, routinely live for many more than one hundred years, and Chicago's Brookfield Zoo, which displays the animals in its reptile house, thinks it likely that a few of the tortoises still living on the islands may well have been around at the time of Charles Darwin's visit in 1835. Not just tortoises but terrestrial and marine turtles may live long lives. But it's good to be aware of John Lawson's remark that fewer than one hundred come to perfection. Caught though it is between decks, the turtle's successful reproductive strategy is that of the fish or the dandelion, which seek the safety of the species in sheer numbers of eggs or seeds, so many, in fact, that the loss of 1,000 allows the survival of one. A marked box turtle may hold the longevity record for its species: 138 years. Which brings me to Crash.

He was discovered marching with slow, steady chelonian determination along a narrow, tree-shaded street in my Shenandoah Valley town. With discovery, his life—and the life of his captor—changed in an instant from Zero-Zero neutrality to Plus-Zero commensalism. And, since 1995, he has spent his days and nights in various Midwestern states—Wisconsin, Ohio, and Michigan, to be exact—far from the Mid-Atlantic state of his nativity. His name arises from the accident, the crash, that he once suffered. It's clear that something knocked him for a loop, injuring him either on the first impact or on a second, when he made a hard landing after flying through the air. The evidence consists of a three-inch gash that runs across his shell near its crest like a miniature canyon. By the time he was taken captive, it had long been healed. He had also, at an earlier point in his life, been interfered with by humankind, for he wore a slathering of dark green paint—house paint, his keeper thinks—the last vestiges of which are only now wearing off His age is uncertain. What was certain from the outset is that he is male, for his eyes are a fiery red, not the docile yellowy brown of a female, and his hinged plastron is concave, while hers is flat (probably to make more room inside for her eggs).

He was joined two years later by Splash and Carolina, both found strolling near the North Carolina coast. Not long after she was taken into captivity, Carolina laid three eggs, none of which produced another of her kind. Each turtle has been ensconced in a separate aquarium, furnished with drinking water and lined in the beginning with wood chips that have now composted into dark brown humus. As this kind of turtle ages, it becomes almost solely a vegetarian, relishing the flesh of fruits, banana peels, and green vegetables, though an occasional worm is welcome. It helps if the food has color-appeal, for turtles have sharp color vision. They are not like cats, keenly attuned to movement and odor, but rather need to see their immobile food in order to approach and eat it. If Crash, Splash, and Carolina ignore what their keeper has put into their dens, she picks them up and places them so that their suppers are visible. Kept perpetually inside, they do not hibernate, though pet turtles kept in outside enclosures in colder climates will indeed excavate winter resting places, going to ground when the air gets nippy and emerging in spring. The keeper of Crash, Splash, and Carolina says that lack of hibernation may shorten their lives. Given their robust survival so far and their potential for hanging on and on and on, it's possible that she herself may not live to find out.

All three are Eastern box turtles, *Terrapene carolina carolina*, the "Carolina terrapin." The reduplication of *carolina* in the name distinguishes these turtles from the other four subspecies with whom they share the name—*T. c. major*, *T. c. bauri*, and *T. c. triunguis*, which translate as "big terrapin," "Baur's terrapin," and "three-toed terrapin." The common names for these subspecies are, respectively, the Gulf Coast, Florida, and three-toed box turtle, which has—as its name indicates—three toes on its hind feet, unlike the others, which have four. Two other box turtle subspecies live in Mexico. A closely related turtle is found in the Great Plains and upper Midwest—the ornate box turtle, *T. ornata*, with elegant yellow dashes decorating the sides and crest of its dark brown or black carapace. Wisconsin lists this species as endangered. The Eastern box turtle is considered endangered in Maine and of special concern in Massachusetts. Why? The reasons are manifold and complex. Habitat destruction, with woodlands yielding not only to agriculture but also to housing and commercial developments. Poaching by individuals and pet suppliers. Extremely high hatching mortality. But even where a species is not pronounced as rare and endangered, it is still part of a wild population. Laws across the country generally prohibit the taking of any wild animal. The federal government provides turtles with an exception, which is sometimes trumped by state laws. Peculiar though it may seem, the Food and Drug Administration is the agency that has set the basic standard that specifies which wild turtles may or may not be kept. The reason for the FDA's jurisdiction will be given in due time.

So, Crash, Splash, and Carolina are held illegally. Nonetheless, their keeper says pointedly that state law bars her from releasing them in Michigan. The reason? Though they are—or once were—wild turtles, not one of them was hatched in that state. In order to turn them loose without incurring a penalty in Michigan, she would be obliged to take them back to their states of origin, which might then charge her with the offense of releasing turtles that could transmit pathogens to wild populations. Damned if she does, damned if she doesn't. She is not, however, worried that the Turtle Police will come knocking on her door.

Kay is. She and her husband ask that I give her a pseudonym so that the TP will not be alerted. From 1991 to 2001, she was in illegal possession of Hambone, an Eastern painted turtle, *Chrysemys picta*, the "painted gold-turtle," which is, incidentally, the official state turtle of Hambone's native Illinois. The species' carapace is plain on top but its underside shows an intricately fashioned mosaic of red, gold, and black markings on the marginal scutes, and the red-brown and light-brown center of the plastron is surrounded by glorious red and gold. Hambone, then the size of a fifty-cent piece, was found swimming in one of the ponds at the local landfill, where Kay's cousin had a side job clearing away troublesome turtles. The little turtle was so new to the world that his carapace was still soft. In the beginning, he flat-out refused to eat commercial turtle food. Luckily, he relished earthworms and learned to eat them out of Kay's hand. Kay says, "He also liked flies and small moths. June bugs scared him the first year—he would dart away from them, then timidly try to bite them fighting them off when they would try to grab his head with their strong legs, and finally he would eat them." Later, she got guppies for his tank; he snacked on the newborns at first but soon became able to catch the adults. Sometime in his first year, feeding him became easier because he decided that store-bought turtle food was acceptable, after all.

In 1992, Hambone was joined by Lilypad, a red-eared slider so named because she looked like a miniature water-lily leaf, no bigger than a quarter, floating on the surface of the same landfill pond. Kay says that Hambone was delighted to have a companion. In three years, his carapace length grew to three inches, and he attained sexual maturity. At that time, Lily, though a full year younger, had grown even larger. Just the same, he lusted after Lily. Kay tells what happened next. "By the following year, she resisted his unwanted advances by stepping on his shell and holding him down until he nearly drowned. When that did not discourage him, she started biting his toes. After she broke one of his toes and bit a second off, I separated them." Hambone could indeed have drowned; fluid in a turtle's lungs is as lethal as fluid in those of a person. In exile, Hambone could still see Lily and for his remaining years got as close to her as possible.

"He was sweet-tempered," Kay says, "and liked to be held and petted. He would paddle upright out of the water following a finger with or without food." On the other hand, Lily is grumpy and does not want to be touched. She nearly took a chunk out of Kay's finger two years ago when Kay was cleaning her. "Nevertheless, she watches us when we're in the kitchen, and when she's hungry, she paddles against the glass of her aquarium. If that does not result in food, she bangs her shell against the glass."

Both of Kay's turtles have been well cared for. Why would the Turtle Police be interested? Because she broke the so-called "4-Inch Law" when she took them in and gave them a home. The law is not a statute passed by a legislative process but rather a regulation first promulgated by the FDA in 1975 and since amended. It covers "all animals commonly known as turtles, tortoises, terrapins, and all other animals of the order Testudinata, class Reptilia, except marine species." The last are excepted because people do not generally keep sea turtles as pets (though hawksbill turtles furnish tortoise shell for combs and ornaments and some people find green turtle eggs irresistibly delicious). The regulation stipulates that, with some exceptions, "viable turtle eggs and live turtles with a carapace length of less than four inches shall not be sold, held for sale, or offered for any other type of commercial or public distribution." The exceptions are the sale of turtles and their eggs for scientific and educational purposes, sale and distribution among private parties (not businesses), and the export of turtles and eggs. It would seem that the regulation applies only when money is involved in a commercial enterprise, but it has been extended in practice to anyone who has brought a smaller turtle inside to keep as a pet or has scooped up a wild turtle and given it to a friend for free. If the TP come knocking, the first thing that they do is demand that the turtle or eggs be destroyed "in a humane manner by or under the supervision of an officer or employee of the Food and Drug Administration." The owner of the undersized turtle has ten days in which to comply. Should the animal not be properly done in by the owner, the TP will take charge, confiscating and killing it—or them, as the case may be. The owner risks being hauled into court, fined not more than $1,000, and sentenced to not more than one year in the pokey. Draconian measures, nor would a bereft owner be consoled by knowing what possessed the FDA to issue such regulations in the first place. The reason is invisible: salmonella. And a rumor accompanies the reason: a tiny turtle might just be popped into a toddler's mouth, and salmonellosis would ensue. Turtles are notorious carriers of those sneaky bacteria, and anyone who handles any kind of turtle is well advised to wash hands immediately afterward. I'm told that cases of salmonella poisoning among turtle-wranglers did drop after 1975. Nor is human health the only consideration. Turtles in captivity may pick up contagious pathogens that could spread among their kind if they were returned to the outside world.

The red-eared slider, once the most prominent ten-cent store chelonian, is the species that triggered promulgation of the 4-Inch Law. And, as winter follows autumn, consequences have followed the regulation's institution. *The Vivarium Magazine*, subtitled *Human/Reptile Interface at the Edge of the Third Millennium*, writes, "An unintended consequence of the four-inch law has been collecting pressure on adult turtles, such as box turtles, and discouraging the development of turtle herpeculture for the U.S. market on a commercial scale." As a result, the variety- and pet-store trade in hatchlings has dropped to nil. Yet, no turtle attains the requisite four inches without having first been a tiny hatchling, and for many turtles, reaching four inches can take years. Some species—Cagle's map turtle is one—never get that big. Yet, perfectly ordinary people want pet turtles. Taking red-eared sliders from the wild would not make a dent in their U.S. population, but other species, like the bog turtle and the spotted turtle, already hover on the brink of endangerment.

Another consequence is the export of turtles. Turtle-farming in the U.S. has by no means come to a halt because of the 4-Inch Law. Red-ears, in particular are still being farm-raised here in uncountable numbers for foreign trade. All by itself, that species accounts for a good (or bad, depending on your point of view) eighty percent of the millions of native reptiles shipped yearly out of the U.S. Where do the red-ears go? To Mexico, Europe, and the Far East. In Japan, they are popular as pets. They have become the second most common turtle in Taiwan. China now farm-raises them, and in 2006, observers saw thousands of them daily in Chinese food and pet markets. The European Union banned the import of red-ears in 1997 because they may have posed a threat to native freshwater turtles, though no documentation of such an impact exists. New World

red-ears have nonetheless made themselves thoroughly at home in the Old World. Nobody really knows what ecological damage a feral population of red-eared sliders might have in lands where they are not native. One sure fact is that native turtle populations abroad have suffered declines in urban and agricultural areas, but it's not clear that the presence of exotic red-ears is in any way to blame. For one, development has diminished habitat. Then, I suspect that the burgeoning of human populations has led to an equal burgeoning in the appetite department. It may well be that red-eared sliders turn out to be the turtle species that conquers the world.

It's not the FDA regulation alone that affects the selling, finding, and holding of turtles as pets. Each one of the fifty states can put the Turtle Police on patrol with laws that restrict the taking and keeping of any wildlife. And the laws are a hodgepodge. In Maryland, you may keep one box turtle and get a permit to have more than one from the Department of Natural Resources. State law also allows the possession of a turtle under four inches, but the DNR has not yet seen fit to honor the law. In New Jersey, you may keep a pet turtle if you get a permit and also provide proof that the turtle was not wild-caught in the state. Indiana forbids the taking of Eastern box turtles from the wild; if you should acquire one legally from another state, you must arrange to have it inspected by a veterinarian and get a Special Purpose Turtle Possession Permit (that's the name that the Indiana Division of Fish and Wildlife came up with) before you may bring it home. Indiana also requires that you obtain a permit to keep a shell that no longer houses a box turtle, and it prohibits the breeding of captive turtles. To take snapping turtles—a legal act—you need a hunting license. Ohio allows you to take a turtle from the wild but stipulates that when its carapace reaches four inches a PIT—passive integrated transponder—microchip be implanted in the animal's skin for identification. In Georgia, it's unlawful, period, to keep any box turtle, no matter how you acquired it. Virginia, on the other hand, is liberal, requiring no permit but limiting the number that may be kept to five Eastern box turtles and five of any other species. All, however, must measure four or more inches.

*

Wumpus measures way more than four inches. And when I hear about Wumpus, I make arrangements to visit her at Suzanne's home near Baltimore. Suzanne, tall, slender, clad in jeans and a pink turtleneck sweater (what else?), has been a turtle-lover since she was a child. Her father and grandfather grew roses in raised beds in several large greenhouses. To control earthworm and bug populations, they gathered box turtles, which were abundant, and kept them in the greenhouses. In the evening after her own supper, Suzanne would take leftover food to the twenty or so turtles gathered in a corner where her grandfather had built a concrete watering hole. She says, "Those box turtles gave me a wonderful education about their kind. I learned what they liked to eat, watched them mate, and then saw how the female started crawling, knocking the male on his back with his hind feet caught between her carapace and plastron. I learned, too, that most, but not all, box turtles can swim." She smiles, remembering her hours with those turtles.

I hear about them a bit later, for she introduces me to Wumpus only a moment after I arrive. Wumpus is completely licit, for there was no 4-Inch Law in effect when Suzanne bought her as a quarter-sized hatchling, still wearing her egg-tooth, in July 1969, at a Korvette's discount department store. She now weighs a good ten pounds and measures almost twelve inches in carapace length. She's as big as a female red-eared slider ever gets and old enough so that her once bright red "ear"—really a comma-shaped patch behind her jaw—has become a dark reddish brown, visible only if you do considerable squinting. How does Suzanne know that Wumpus is female? Not only are her foreclaws stubby, but she also lays eggs. Since attaining sexual maturity when she was seven or so, she has laid six or more clutches every year. Nor is there a season to egg laying; it may occur at any time except for winter months.

Laying eggs used to be an outdoor event, calling for much effort on the part of both turtle and woman. Because the process took time, Suzanne lugged a chair, plus magazines, binoculars, and camera, as she followed Wumpus while the turtle plodded purpose-driven up a winding, graveled driveway almost to the road. One reason for following her was that Wumpus understands nothing about cars. Another reason was that it's thrilling to watch Wumpus as she goes about her work. She would make several trial excavations in the soil at the edge of the road before she dug one that really suited her. Sometimes, she'd decide not to lay after all. Nowadays, Wumpus plops out eggs when she's ensconced in her tank. Suzanne says that most of the time she lays only three or four eggs but, a week before Easter 2007, she produced a full dozen. After they are laid, Suzanne collects the eggs and puts them outdoors as food for raccoons and possums.

While I visit, Suzanne brings her into the living room. Wumpus knows where she is and, for the most part, moves easily around the house, though she has on occasion gone thumpity-bump down the stairs to a lower level of this dwelling built on a hill. When she wishes to be outside, she goes to the sliding door in the living room and indicates her desire by scratching on the glass. Suzanne goes with her to make sure that she stays safe. It's not problematic to let Wumpus roam indoors, for she defecates immediately after she eats, and she never eats anywhere but in her water-filled tank. When Wumpus is ready, she will return to her tank in the kitchen and thump against it until someone lifts her into it. As with Kay's Lilypad, shell-thumping is a prime means of communication. She likes to have her carapace scratched; if she didn't like it, she would move away right smartly, but as it happens, she goes into spasms of delight, like a child being tickled. Her four limbs thrust out rapidly out, then just as rapidly retract. After that, a queen surveying her domain, she strolls around the room before she stomps, thud, thud, back to the kitchen. I lift her into her tank, which is a large plastic basin about twenty-nine inches long, twenty-one inches wide, and about foot high. Filled halfway with water, it is lit by a full-spectrum light and kept close to eighty degrees. It is clean, clean, clean, and so is Wumpus. Suzanne scrubs her with a brush to remove algae one to three times a day—every time the tank is emptied and refilled. Wumpus regularly sheds her scutes—the keratinous coverings of her shell; they are patterned with brownish patches on a pale background. Suzanne collects them and says that some of the shed scutes easily exceed Wumpus's size as a hatchling. And what does Wumpus eat? "Earthworms in summer, and any outside bugs," says Suzanne. "Romaine lettuce, berries, other veggies, and commercial turtle food. The little guys get Turtle Brittle."

The little guys, oh yes. Wumpus is not the sole turtle on the premises. The living room holds three other tanks, all smaller than Wumpus's, but then the inhabitants are not nearly so regal in diameter. Each tank (originally, a sweater box made of rigid, heavy-duty plastic) is lit with a full-spectrum light; each is kept at eighty degrees by a heating pad laid underneath; and each is furnished with a turtle house made of a Styrofoam container with a doorway cut into it. One of the tanks provides a habitat for three box turtles between two and two and a half inches in carapace length; another holds a tiny, new-hatched box turtle; and the third, an equally tiny painted turtle. Why flout the 4-Inch Law and court not just confiscation but also destruction by the Turtle Police? Because these found turtles would almost surely be destined to become dinner for a raccoon, a crow, or some other carnivore much larger than an infant turtle. Suzanne will raise them until they are big enough to survive on their own, then return each one to the very spot from which it was taken. Even these little reptiles are distinctly different, one from another. Rex, the largest of the box turtles, will emerge from his house when he hears Suzanne's voice. "All animals have personalities," Suzanne says. "The little ones are so funny. One has commercial Turtle Brittle augmented with calcium to provide vitamin D3, which turtles need as much as people do for promoting bone strength. I am enchanted and pick up the painted turtle to look at its beautifully embellished underside. It waves its limbs frantically in the air: Put me down, put me down! So I do—and go to wash my hands."

The next morning, Tommie, a dedicated herpetophile, arrives in an ancient white Ford pickup truck to join us for breakfast. He wears his grizzled hair in a ponytail and sports an equally grizzled, neatly trimmed mustache and beard. His blue eyes blaze with a passion for chelonians of any time and any place as he shows some of the fossil shells that he has collected. A totally enamored turtle fan from the word go, he tells me that when he was eight years old, he would push the canned terrapin and turtle soups to the back of the grocery-store shelf on the theory that if none were sold, the store would stop carrying the stuff. He's a storehouse of turtle miscellany, explaining, for one thing, that turtles pile up on slanted logs because they like to have their shells at a particular angle to the sun, and piling assures them of getting the right angle. For another, tortoises, like freshwater turtles, communicate by butting with their shells. "They think you have a shell, too," Tommie says, "but you're just a dumpling." And he tells the tale of the seventy-pound Burmese mountain tortoise that wedged itself under a vending machine in the Baltimore-Washington International Airport. The vending machine was nearly flush with the floor. How to retrieve the beast? The tortoise itself solved the problem by using its sturdy legs like hydraulic jacks and lifting that heavy machine up, up, up into the air, whereupon Tommie and a colleague could tilt it back and recover the escapee.

And why was sizeable Burmese mountain tortoise at BWI in the first place? Because Tommie engages in international chelonian rescues, and for the same reason that Suzanne does so domestically: to keep the animals from being eaten, in this case by human beings. Through what Tommie calls "the turtle grapevine," he learned of several Burmese mountain tortoises (*Manouria emys phayrei*) being held in a Malaysian food market. Their destiny: the soup pot—until Tommie came to the rescue, buying them and arranging for transport to the U.S., from Malaysia to Los Angeles and from there to BWI. He keeps them in his basement, which he describes as smelling like a ripe cow barn.

The Burmese mountain tortoise, sometimes called the Burmese forest tortoise, is a peculiar beast that shuns arid regions in favor of moist forests in monsoon country. It is native to easternmost India and Bangladesh through Myanmar (which used to be known as Burma) to western Thailand. And it is the largest tortoise found in Asia. Its genus name, *Manouria*, was invented by an Englishman, John Edward Gray (1800–1875), and "invented" is surely apt, for the name seems fancifully devised. A French journal, *Revue francophone d'étude, d'élevage, et de conservation des chélioniens* (*French Review of the Study, Breeding, and Conservation of Chelonians*) says this (I translate):

> If the name Manouria seems instilled with poetry and sings sweetly on the ear (of the informed turtle-lover), the history of the nomenclature of this genus and species is very obscure and refers to nothing epic. It's probably futile to look for a coherent etymology, the more because Gray had at hand only

a single carapace.

I think that Gray was having fun. He may also have come closer to a living animal—or have heard about it from someone who knew it well—than the article allows. The article follows up its comment about the single carapace by saying that English has only two words that are anywhere close to the tortoise's genus name: manorial and manure. Tommie's description of his basement's odor comes immediately to mind. The meanings of the species and subspecies names are not so fanciful. *Emys* is Greek for "tortoise," and *phayrei* honors Sir Arthur Purves Phayre (1812–1885), who was appointed commissioner for the entire province of British Burma in 1862. In 1883, he published the first standard history of Burma. One enduring result is that a host of animals bear his name as their species designation, such as the leaf-monkey, *Trachypithecus phayrei*, "Phayre's shaggy ape," and a squirrel, *Callosciurus phayrei*, "Phayre's thick-skinned squirrel." But it's not Sir Arthur who deserves attention; it's the tortoise itself.

This stolid creature is believed to be one of the most primitive representatives of tortoisekind. Among its antique features are its broad, flattened shell and its preference for cool, moist forests. In captivity, it needs a natural, tree-shaded enclosure furnished with mud wallows or the humid environment of a greenhouse. The behavior of a female mountain tortoise when she intends to perpetuate her kind is unlike that of any other tortoise. Her instincts propel her to build a nest-mound of leaf litter that sometimes reaches two and a half feet in height and a full five feet in width. Just before she lays her eggs, she goes into the mound headfirst to scrape a depression in the earth below the leaf litter. When she emerges, she places her cloaca over the depression and proceeds to lay a clutch of eggs more numerous than that produced by any other kind of tortoise. Nor does she simply stroll away when she's emptied herself. Instead, she stays beside the nest to guard it. She'll push away nosy animals with the front of her shell, and if that doesn't work, she'll climb atop the mound and spraddle herself across it with limbs fully extended. But after two or three days, she's had enough and lumbers off. Given a mate and a similarly moist, leafy environment, she'll breed successfully in captivity. And it may be the species' salvation, for the human pressures on it in its native lands are immense and ravenous. Its numbers diminishing, this tortoise's status is of special concern. It's easy to understand Tommie's drive to rescue it.

*

My chelonian octet—George, Timothy, Crash, Splash, Carolina, Hambone, Lilypad, and Wumpus—began their careers as tiny, fragile, easy-to-catch newborns. Turtlekind's inborn knowledge decrees that the more hatchlings, the better the chances for the survival of a lucky few. Four of the octet were taken from the wild as adults, while the others—two of them purchased, two caught in the wild—were still minuscule when they became pets. How many died as the eight of them were growing up? It's safe to say that John Lawson's theory of a hundred to one is vastly understated. A rule for pet turtles, be they store-bought or captured, is that their owners can expect many more of them to die than to live. Today, people are crowding out and eating up the feral populations of these reptiles to an extent unknown in earlier times. But the 4-Inch Law provides no relief at all. And it's outdated (if ever it was in date). It is, moreover, applied capriciously. Then, because the FDA regulation and state laws have adverse effects on wild populations of grown, reproducing turtles, and because they restrict scientific research, Tommie and many other turtle aficionados support repeal. So does *The Vivarium Magazine* with one exception, that it be illegal to sell turtles to minors. I believe that the reason for that exception is not that minors might put turtles in their mouths but rather that they are less likely than adults to care for turtles properly. And a turtle's needs for attention are almost endless, as Wumpus demonstrates. It's possible, though, that red-ears may return to pet stores, for the U.S. Senate has approved repeal of the 4-Inch Law and, as I write, it will soon be up for debate in the House. But there are several catches to repeal. One is that the hatchlings be certified salmonella-free. In Louisiana, turtle farmers, who've made a living for the past thirty-plus years by raising turtles for export, have been working with scientists at Louisiana State University to develop methods that eliminate salmonella in their stock. The process is complex; eggs are given a bath in water that contains bleach; and, afterward, loaded into a specially built egg-cleaning machine. The result: a reduction of salmonella contamination to less than one percent. The problem is that recontamination of squeaky-clean hatchlings can occur in an instant because of environmental factors, like being fed hamburger or sharing a tank with a big wild-caught turtle. The second catch is that federal repeal would leave state laws firmly in place, and the regulations on who may and may not own turtles of any size would remain as capricious as ever.

If little red-ears do return to the American market, providing a care-booklet with each sale might be made mandatory (though people are often as reluctant to read such stuff as they are car-maintenance manuals and the warning labels on ladders). We have domestic turtle farms and dealers, to be sure, but they do not—cannot—wait for hatchlings to get big enough to sell as pets but instead send the little reptiles to the ends of the earth. Would-be pet owners are mostly out of luck, even though information on the care and cosseting of turtles is much improved. "Mostly" because adoption programs are available through the many tortoise and turtle societies founded by their ardent fans. These societies also provide practical advice, as well as sponsoring scientific research and conservation efforts for threatened and endangered species.

A warning needs to be passed on to people who keep turtles. And it needs a paragraph of its own. Though the Rev. Mr. Gilbert White thought that Providence had made a mistake by blessing a stodgy reptile with longevity, your turtle may well live longer than you do. Make arrangements for its future care and feeding before you depart this world.

More power to Suzanne and Tommie, to Kay and to Crash's keeper. The future of turtlekind may lie in their hands.

EIGHT

Sparrow Crimes

{ A Story of Immigration }

A small, male sharp-shinned hawk perches on the foot-high concrete wall that separates my front yard from the sidewalk. With a long, banded tail and a white breast barred with rust, he is handsome. His eye seems to be set on the froth of golden yellow flowers covering the large forsythia bush at the edge of the yard next door, but he knows what the flowers conceal, and his body is as still as the concrete wall. The stillness is

broken in a flash, and in a second, the hawk settles back down with a sparrow in his talons. The bush erupts, exploding dozens of sparrows into the air like feathered confetti. One of their number had become unwary enough to expose itself. Its life has bought a brief respite from predation for the rest.

The scene repeats itself frequently when the forsythia is in flower or in leaf. Though this little, native hawk—*Accipiter striatus*, or "striated swift-flier"—is characterized by one of my bird books as "a shy, secretive woodland hawk," it catches its dinner wherever the ingredients—tasty small birds—are handy. If doing so means perching in plain sight, so be it. Appetite wins out over exposure. And, if need be, the sharpie can put on a burst of speed.

The sparrow that I see caught and the sparrows that burst from the bush are another matter altogether. Their stout-billed species is formally known as *Passer domesticus*, which translates directly into their common name "house sparrow." The binomial was given to them in 1758 by Carl Linnaeus, who certainly outdid Adam in the business of naming living things. And the birds are alien, a fact attested to by another of their common names—English sparrow. England, however, is only one of the places in which they are indigenous; the species spans Europe, North Africa, and Asia, and has been introduced elsewhere around the world. With the exception of the antipodes, they may be found everywhere. Since 1851, when a British flock of 100 was released in Brooklyn, they have made themselves thoroughly at home all over the United States. (Some sources will tell you that they were brought to Central Park, but that was the site at which the European starling was introduced.) Other flocks were brought subsequently to several other cities—Salt Lake City and San Francisco, for two—in the late 1800s as a means for pest control.

Today, house sparrows reign from coast to coast, as well as north well into Canada and south as far as Panama. Recent estimates put their population in the U.S. alone at between 150 to 400 million, though it seems that their multitudes are, for reasons unknown, declining slightly. It may be that new practices in agriculture, such as the single-crop focus of many factory farms, have reduced the availability of favored foods. Constructing houses and outbuildings with fewer nooks and crannies may also play a role by denying nest sites. Their population, nonetheless, is the third greatest, after mourning doves and starlings, of any avian species living in the U.S.

But the American decline in house sparrow numbers is noticed, I'm sure, only by those who make bird censuses. The sparrows themselves do all they can to keep their numbers not just stable but on the rise. In breeding dress, the males are attractive even to human eyes; they wear a black bib and a soft gray, chestnut-bordered cap. The lightly streaked females define the acronym LBJ—Little Brown Job. Each monogamous pair raises three or more broods a year, and each brood may consist of five or six nestlings. The incubation period of only ten to twelve days is the shortest of any bird.

But house sparrows to control pests? Hardly! Insects comprise a paltry four percent of their diet. Most of the bugs that they catch are given as food to their nestlings. They themselves have figured as pests from the moment of their arrival in the New World, for seeds are their primary food, and their appetite for the grains of cultivated crops is bottomless. In the early years of their residence in the New World, they even learned to use the undigested seeds found in horse manure. Then, they evict native birds—notably, bluebirds, tree swallows, and martins—from their nests by tossing out the furnishings and sometimes smashing the eggs. The reason for such violence is that house sparrows, like their victims, are cavity-nesters, though they will deign to use soffits and, on occasion, bushes and trees. Gregarious creatures, congregating by the chirping hundreds, they have become as much a city bird as the pigeon. They can be seen in almost any sort of urban venue—on downtown sidewalks and streets, under the overhang at a gas station, a movie marquee, or, as poet R. T. Smith puts it, "the roof / of Lowe's amazing discount / store," where he saw them foraging on spilled birdseed. And when they want a heartier dinner, they're perfectly capable of taking wing and heading for parks or the countryside to find the requisite seeds. Some house sparrows, however, have left behind their outdoor foraging and made themselves permanently at home within walls. I have seen them in the food court of the C Concourse at LaGuardia Airport, where they were dining heartily on the crumbs dropped by sloppy travelers.

Various strategies have been employed to eliminate *P. domesticus*, or, at least, to reduce the damage it wreaks on crops, other birds, and human sensibilities. In the mid-1800s, Britain saw the formation of sparrow clubs that aimed to exterminate the species, and bounties were paid until the late 1800s when it became evident that such efforts were as useless as commanding the tide not to come in. Today, killing house sparrows is illegal in the United Kingdom, and to that end, contractors repairing buildings must make certain that the crannies contain no nests. The status of the house sparrow in the United States represents the opposite view: these sparrows are one of only three species—the others are pigeons and starlings—not protected by the statutes that make the killing of migratory birds a crime. But, lest they seem fair game, be cautioned: state laws apply, and a hunting license is required for those who'd pot-shot at the bird. Unlike game birds, however, there is no bag limit.

I think that it's mightily improbable that most people are aware of such laws; but even if they are, they'll take their chances, for house sparrows, all unwitting, are capable of arousing murderous impulses in lots of folks, including my father and one of my brothers, who committed their sparrow-crimes on the family farm not far from the banks of Lake Erie.

Some stroke of fortune has spared my father's journals, which he kept faithfully from his twelfth year through his twentieth. His neat handwriting covers the day-by-day pages of inch-thick, clothbound journals issued by the Cleveland Real Estate Board. Many of the entries are innocuous, like this one from October 12, 1922, when he was thirteen: "Cleveland, Ohio. I went to school. I went to bed. It haled real hard this afternoon." And this harder-hitting entry on August 3, 1923: "PRESIDENT HARDING DIED LAST NIGHT. He died about 10:30 last night. We were coming from Buffalo on the train. We were eating breakfast when the Steward brought us a paper telling of it. Vice-President Coolidge was sworn in at about 2:30 this morning." The first indication that he considered birds fair game occurs later that month, on the twenty-fifth: "Dad brought me a .22 Stevens gun 'Favorite.' I shot 2 pidgeons. It sure is a dandy gun." Two days later, he shot three more.

It is in 1924 that he begins to tally not just pigeons but also house sparrows. The first mention occurs on Saturday, May 10: "I went to the farm in the afternoon. I got about 50 sparrow eggs. I fed them to the chickens." Thereafter, the journal shows a mounting tally. Many summertime entries consist of not much more than numbers:

Sunday, August 3:

147
13 more sparrows
160

A trap was his main means of dispatching his prey, but he occasionally employed a farm dog and a firearm to do them in: "Wednesday, August 27: In the morning I raided the little chicken coop. Sheppy was with me. He killed 7. I made a second raid netting 2, a third netting 9 and a fourth netting 1. Caught 7 in trap." His total for that day was twenty-eight, the second highest number for that year. The total for 1924 amounted to 351 sparrows. The number almost tripled in 1925 to 923, with an additional 200 chalked up for pigeons—more than a thousand birds dispatched summarily to kingdom come. No killings are recorded for 1926, but then, he'd fallen in love, and his attention was focused on the woman who was, seven years afterward, to become my mother. Much later, on the same farm, my brother always aimed at house sparrows with his .22. It was loaded with dustshot, a shotgun shell packed with tiny pellets and miniaturized to fit into a rifle. He'd sit in an outbuilding, watch a feeder hung nearby, then pot the feeding sparrows one after another. "Of course, I wouldn't do that now," he says, but his face is happily suffused with memories of great success.

Is anything about house sparrows lovable? House sparrows have served famously as pets. More than two thousand years ago, the Roman poet Catullus (84–53 B.C.) wrote, *Lugete, o Veneres Cupidinesque . . . passer mortuus est meae puellae*:

Grieve, lament, all arousals, all
passionate desires . . .
my girl's little sparrow is dead,
the sparrow, my girl's delight,
that she loved more than her own eyes

And grief has made those eyes red and swollen. Not only has death claimed the sparrow, it has also taken away his girl's beauty and her sexual allure. Weeping and wailing also assail Jane Scroupe, a pupil of the Black Nuns of Carow, to whom John Skelton (c. 1460–1529) addressed his poem "Phyllyp Sparowe." Though the orthography of the bird's given name wanders throughout the poem in not-unusual sixteenth-century fashion, the name itself is meant to replicate the sparrow's call, most often rendered today as *chirrup*. Poor Jane Scroupe! Bereft, she cries:

When I remember agayn
How mi Philyp was slayn,

Never halfe the payne
Was betwene you twayne,
Pyramus and Thesbe,
As befell to me:
I wept and I wayled;
The tearys downe hayled;
But nothing it avayled
To call Phylyp agayne,
Whom Gyb our cat hath slayne.

Jane recalls the sparrow sitting on her lap seeking white bread crumbs or lying and napping between her soft, bare breasts. To Philip's funeral she invites all the birds she can think of, including the Phoenix, and she prays for Philip's soul. I rejoice in her recognition that a bird has—not can have, but *has*—a soul.

So, house sparrows have indeed been found lovable by some. And I can imagine caging a pair of house sparrows as one would cage parakeets or canaries. It would have to be a pair, at least, for the birds are highly gregarious. As for keeping a sparrow for a pet that feeds from a hand or snuggles against one's skin, some patient person might well succeed. I've heard of an indoor house sparrow that its keepers named Busy Bird, and I know people who feed chickadees and squirrels with tidbits held on their palms, but the world today moves faster than it did in the times of Catullus and Skelton, and we have diversions too numerous to count. The birds, though, are not unattractive. In 1999, the Faroe Islands saw fit to honor the species with a postage stamp featuring an LBJ and her mate in breeding dress.

I find the speed with which they adapted themselves to North America certainly deserving of admiration. In fact, house sparrows are now as American as the European starling and kudzu, as American, indeed, as anyone descended from African slaves, Ellis Island immigrants, Cuban boat people, or those who crossed the land-bridge over the Bering Strait some eleven millennia ago. We can't send *P. domesticus* back where it came from. It merits acceptance if not an outright welcome.

NINE

Melanodon's Children

{*The Survivors' Story*}

Something streaks across the kitchen floor so fast that it is only a small gray blur, like a shooting star of very low wattage. The Chief and I cannot see head nor tail nor scurrying legs. But we know what it is: a common house mouse. We don't, however, know what has impelled it to cross the kitchen at great speed on a fairly regular basis. Mousetraps are bought and set with peanut butter. Two mice are quickly dispatched. The mad dashes stop. Only a year later do I discover the reason for them when I clean a bookcase around the corner in the dining room. Behind the books on the bottom shelf lies a small mountain of dog food. The daring twosome were raiding the dog's dish and carrying away chunks of kibble as almost big as their heads.

I've lived in my Virginia house for twenty-four years. Before the dog-food raids and afterward, no signs at all of mice. I am surprised, though, that the streaking creatures did not leave their calling cards, those little fecal pellets that look for all the world like black grains of rice. In our North Carolina mobile home, a porous dwelling, we'd frequently find cotton mice that had abandoned the outdoors most likely for the sake of food. I'd wake in the morning to discover that they had been dancing in the kitchen during the night. Out came the traps, and after two or three or five had been caught, there'd be a respite, usually short. Dancing on the kitchen counters was not their only game. They built nests in dresser drawers, sometimes using the pink fiberglass insulation from the electric stove. Once, a nest was constructed of red wool from an intricately cabled cardigan that had taken me months to knit. I bagged much unraveled yarn and the tiny pink babies cradled in it and put the whole works in the trash. At that point, I could gladly have trashed every last mouse that had ever lived. Instead, I bought more yarn and stored the new sweater in a heavy-duty, zippered plastic bag.

But one small invader brought on an event that amazed me. We always disposed of the little carcasses in the river that flowed seventy-five feet away from our door. One morning just after dawn, I walked to the river carrying the night's catch by its tail and tossed it in. Back inside, breakfast on my mind, I happened to look out of the kitchen window. Oh my! Standing at river's edge, a great blue heron was busily

tossing the mouse into the proper headfirst position for swallowing. The bird had retrieved it from the water. I was not the only one with breakfast in mind. And it was a matter for rejoicing that the mouse could be so handily recycled.

*

The ancestral mouse danced in the dinosaur days. About 205 million years ago, in the Upper Triassic, small, furry, mouse-sized creatures gave birth to living young and nursed them. The mammals had even that long ago begun to appear. And where had they come from? From the dinosaurs themselves, in particular, from the therapsids, a group of lively, pint-sized dinosaurs that were given, not unlike mice, to rapid movements on all fours and much quick running around as they looked for food and mates. The name of this group comes from the Greek words for wild animal and arch, the latter because the animal had developed a stronger arched palate. The new and better palate would be essential for mammals so that an infant would have a sturdy upper counterpart to its tongue as it suckled. (I see that creature: pink, perfectly hairless, and blind, but pulling lustily at its mother's teat, just like a baby mouse.). The transformation of reptile into mammal took place a little at a time during the entire 40 million years of the Triassic. Scientists trace this evolution through changes in skulls and, especially, in teeth. The reason for using those parts as criteria is that not much else survives in the fossil record. Teeth, however, are the hardest tissue in the body and so are able to resist eons of wear. The ways in which teeth are structured makes it possible to determine what kind of food an animal was designed to chew, be it vegetables, meat, or both. Jawbones and teeth underwent two distinctive modifications. One is that the seven bones of the therapsid jaw became one in mammals. The second is that some of those multiple jawbones proceeded to migrate to the mammalian ear. The reptile has a single ear-bone, the columella, which became the stapes, one of the three tiny bones, or ossicles, found in a mammal's ear.

The people who study such things have chosen to give many mammals common and scientific names that are related to their teeth or jaws. The suffix *-don* or *-dont*, from classical Greek, means tooth, while *-gnathus* denotes the jaw. One familiar name, mastodon, means "nipple-tooth" for the nipple-like cusps on the animal's molars. And the common name of the saber-toothed tiger honors its large canine teeth, which easily killed and tore apart prey. Back in the long, slow days of mammalian development, the ancestral therapsids were called cynodonts, "dog teeth," for their specialized dentition which resembled that of today's dogs. And they did look something like heavy-bodied, low-slung canines with massive heads and plump, stubby tails. Likely, they were warm-blooded. Skull bones have been found with little holes marking the site of probable whiskers. And if they had whiskers, they also had hair. One of them, the carnivorous *Cygnathus*, or "dog jaw," has been imagined by scientists as a creature with several narrow stripes running from just behind its ears to the tip of its blunt tail. The teeth provide the clue to its carnivory, for today's meat-eaters have similar dentition. But these animals laid eggs. They had not yet turned all the way into mammals. By the end of the Triassic, therapsids were on their way out, and mammals in many forms had begun their most successful march into the present. Nonetheless, dinosaurs and mammals lived side by side for some 150 million years during the Mesozoic period.

Many of the true mammals were also dubbed with names that described their choppers. The earliest order of mammals is the Morganucodonta, the "Glamorgan-teeth." (Glamorgan is the county in South Wales in which the type fossil was found.) Within the order, the prominent genus was *Eozostrodon*, "dawn girdle-tooth." "Girdle" denotes the large ridges of the animals' upper molars. Some of these creatures were fully as big as shrews, others as small as mice. And they nursed their young. As an anonymous limerick puts it:

Morganucodonts suckled their brood.
They breathed in and out as they chewed.
Their molar facets
Were certainly assets,
But still, locomotion was crude.

As it happens, however, their skeletons suggest that they could not only scurry over the ground at a good clip but also climb upward right smartly. Likely, they were nocturnal and did most of their dancing at night.

The Pantotheria, whose name means "all animals," were the true founders of all three mammalian lines. They lived and thrived in the Middle Jurassic period, about 187 to 163 million years ago. A recreation of one pantotherian species, *Melanodon,* or "black tooth," shows an animal that looks for all the world like a furry, long-nosed, bright-eyed mouse with its bulging cheeks packed full of that era's equivalent of dog food. As for the black teeth, it's reasonable to guess that the fossils were that color, for nomenclators are often best pleased by simple, descriptive terms.

A complex phenomenon known as the K–T boundary, which took place about 65 million years ago, marks the end of the Mesozoic era and the beginning of the Cenozoic. K represents the Cretaceous period, the last part of the Mesozoic; T stands for the Tertiary period. This is the time in which the third largest extinction of life on earth occurred. Two-thirds of the species then living went glimmering forever, among them, famously, the dinosaurs and, much less so, most plankton, many marine invertebrates, and some terrestrial plants. Educated hypotheses have it that the mass die-offs may have been triggered by two huge events, one from outer space, the other from Earth's molten belly. At that time, an asteroid did plummet earthward and crash into what is now Mexico's Yucatán peninsula; the impact sent out shock waves equivalent to a hydrogen bomb blast of 100 million megatons. In the same million-year timeframe, volcanism was at work; thunderously huge eruptions spewed basalt over some 800,000 square miles in the part of the world that we know as the Deccan Plateau in western India. Another reasonable conjecture is that climate change also played a part in extinguishing the dinosaurs and some other life-forms. Summers may well have become hotter, winters more frigid, thereby altering the availability of food. Growing seasons would have been severely discombobulated, and the circulation of nutrient-filled ocean water, just as disrupted. The K–T extinctions may well have come about through a combination of these events, each one exacerbating the others. But we don't know. What we do know is that much did survive: insects, flowering plants, birds, and little mammals, and, in the sea, many fishes, snails, and corals. The proto-mouse had made it through, along with the cockroaches, horseshoe crabs, and red maples. In the Tertiary period, the mammals underwent dramatic diversification and branched in three directions: monotremes, like the egg-laying but milk-supplying duck-billed platypus; marsupials, like the opossum and the kangaroo, whose almost embryonic young are reared in pouches; and placentals, those fed through an umbilical cord during gestation—those, that is to say, with belly-buttons, like giraffes, antelopes, dolphins, lemurs, cats, dogs, and mice, not to mention people. But at the end of the Tertiary, true mice—those that we could not doubt were really mice—had not appeared, not yet.

*

If there are two life-forms that own the world, it may well be the cockroaches and the rodents. The animals with all the equipment, dental and otherwise, that identifies them as true rodents appear in the fossil record starting at the end of the Paleocene epoch some 56 million years ago. Their order, Rodentia, the "Gnawers," includes an omnium-gatherum of creatures with incisors like chisels that continue to grow throughout the creature's entire life. The front of these teeth consists of several layers of enamel, while the back is made of softer dentine. When the latter is worn away by frequent gnawing on whatever is available, from barks and nuts to soap and candles, the former keeps a razor-sharp edge. The rodents are by far the largest group of mammals, representing a full three-eighths of the 4,000 mammal species alive today. And one family within the order, the Muridae, or "mousekind," comprises some 1,100 species all by itself—that's more than a fourth of all mammals. (Them and Us—we are clearly outnumbered.). My streaking, dog-food-stealing house mice and the white-footed mice that un-knitted my sweater are among the smaller members of this enormous extended family. Other members of the family are horridly familiar—the various incarnations of *Rattus*, such as the brown rat and the black rat (a rat by any name is still a rat). Some members figure as pets—hamsters, gerbils, and, sometimes, even mice and rats. Muskrats, voles, and lemmings also inhabit the family. South America, as a separate continent during much of the Cenozoic era, produced an astonishment of Muridae, both large and small. Some of them, like the guinea pig and the fur-bearing nutria and chinchilla, are well known, while others seem invented by mad scientists—the capybara which can grow to weigh more than a hundred pounds, the paca and the tuco-tuco that dig burrows, and the Patagonian cavy, a short-tailed, rabbit-eared speed demon. Nor do rodents stop with mousekind but also include the Sciuridae, "squirrelkind"—gray, red, and fox squirrels, chipmunks, gophers, groundhogs, prairie dogs, and more, most of them with prominent buck teeth. Other rodent families variously contain muskrats, porcupines, and beavers.

The word Muridae comes from *mus*, the Latin word for mouse. The Romans were not nearly so ready as we are to distinguish between kinds of rodents. For them, mouse came in two versions, *mus musculus* and *mus maximus*, "little-mouse mouse" and "biggest mouse." The first, of course, refers to the small mice of house and field, while the second denotes none other than the rat in any of its several species. Today, *Mus musculus* is not just any small mouse but is the name given to the common house mouse, the very small creature that cached a hearty ten pounds of rainy-day supplies behind my books. The cotton mouse that danced by night in my mobile home's kitchen is *Peromyscus gossypinus*. The genus, which is native to North America, came into being some 25 million years ago. It shares a common ancestor with the house mouse and the rat, but the latter two did not become separate genera until 15 million years ago. The cotton mouse has numerous genus-mates, including the deer mouse (*P. maniculatus*) and the white-footed mouse (*P. leucopus*). The meanings of their species names are "cotton," "little-footed," and "white-footed," respectively. As for the generic name, *Pero-* is Latin for boot, and *-myscus* is Greek for "little mouse"—"little mouse in boots"—because the feet and legs of all members of this generally brownish genus are clad in soft white hairs.

The back and flanks of my dancing cotton mice range from an orangey brown to plain brown, with the darkest color on the back; the animals' belly, feet, and the underside of the chin gleam white. Cotton mice aren't picky about where they live, selecting from a variety of habitats—cypress swamps, thickets, piles of vegetation in a clear-cut, and houses. Gestation lasts for twenty-three days, and seventy days after they are born, the helpless, blind pink pups are all grown up and ready to reproduce. The animals also engage in athletics of several sorts, for they are not loath to swim and they can scramble up trees as readily as did Morganucodont. Sometimes, they even build their nests aloft. Mouse in Boots is a Methuselah among mice, for it can live a truly long life of up to eight years, compared with an average span for other genera of four to five years and only two to three for the house mouse. In the wild, however, the lives of cotton mice may not last longer than a year, for they serve as an important part of the diet for hawks, snakes, and other carnivores. In wild and specially bred mutant strains, they also serve as laboratory animals in studies of genetics and evolution.

But it's the relatively short-lived *M. musculus* that has truly made its way into every last crevice of the humanly habitable world. The common house mouse originated in Asia, most probably northern India. Its relationship with humankind is now commensal. But it predates us by millennia. How it got along before people came on the scene is unknown, but an extrapolation may be made: the species is extremely adaptable, able to live in an eclectic assortment of habitats and to nourish itself on a wide variety of plant and insect foods. It was—and is—a truly versatile creature.

Once we arrived on the scene, it found paradise, snuggling right in, helping itself to our food, and making itself comfortably at home in our caves, tents, huts, vegetable patches, and grain fields. By 8000 B.C., it had made its way to the Mediterranean. It did not migrate on its own but rather tagged along when people went on the move, and it rode in wagons, wheelbarrows, travois, rafts, and whatever else people used to transport themselves, their food and their other worldly goods. Often, in warmer weather, it will leave its well-peopled habitat and take a summer vacation in the wild. It doesn't, however, compete well with other rodents and depends for its food and housing, for its very survival, on its association with us. It reached the shores of the New World with the conquistadors and colonists and crossed the Pacific with Captain Cook and other European explorers. Today it's found on every continent except Antarctica and is, along with the rat, the most widespread mammal other than us. Many of us rightly regard the little creature as a pest, for it can contaminate food, destroy the seeds for crops, and spread diseases, from salmonellosis to the plague. More than two thousand years ago, the Roman poet Virgil listed *mus* as one of several scourges that could bedevil a hardworking farmer when it "sets up housekeeping underground and builds its granaries" by stealing the farmer's seeds. Nocturnal creatures for the most part, they are handily omnivorous, feasting on just about anything that they can find, from seeds to bugs. They climb with great agility and when they swim, they are able, it seems, to hold their breath. Accounts exist of mice that have been flushed only to reappear minutes later. And these little animals may build simple tunnels or tunnel systems as complex as those of a groundhog, with multiple chambers and exits.

Yet, the common house mouse leads other lives that have nothing to do with its natural inclinations. It pops up frequently in habitats other than the wild or those that it shares commensally with us. We have several notable subspecies: *M. musculus laboratorius*, the laboratory mouse; *M. m. litterarius*, the literary mouse; and *M. m. deliciolus*, the pet mouse, sometimes called the fancy mouse. Once upon a time, when she was in high school, my younger daughter, Hannah, spent a year raising pet mice. She was ever given to year-long enthusiasms—angelfish, which she not only raised but also bagged and took to tropical fish shows, and plants—dracaenas, palms, philodendrons, a rubber plant, dieffenbachia—which turned her room into a steamy green jungle. As I recall, the mouse year started with six mice in a single aquarium and ended with hundreds—black, white, brown, spotted—in banks of aquaria lined up against every wall. The mice slept in little heaps during the day, but the wheel in each aquarium was in nightlong use, filling the room with a continuous *whirr*. The air acquired a smoldering, musky smell that emanated somewhat from urine and feces but mainly from the pheromones of the male mice, a smell that was sopped up by bed linens, curtains, and clothing. Luckily the door could be kept shut. My household helper refused to enter the place. Though unpleasant to the human nose, the mouse nose uses odors for purposes of recognition. Smells confirm such important matters as the identity of family, friends, and strangers; the location of the home nest; and the presence of a female in heat. The mouse year's final curtain fell when we sold the house and moved. Just how the million mice were disposed of, I no longer recall. Some were surely given away, and others must have been donated to the county's nature museum as snake food. I doubt, though, that any were released to fend for themselves in the great and perilous out-of-doors.

Aside from their pinkie pups, house mice have also spawned a quite a following in the form of clubs for people who keep pet—or, as they'd have it, fancy—mice and rats. Such clubs exist, so far as I've been able to ascertain, in the U.S., Canada, many European countries, South Africa, Australia, and New Zealand. I suspect that keeping mice as pets is a phenomenon restricted to the Western world and its outliers and that elsewhere, they are regarded as destructive, pestilential plagues. The clubs—among them, The North American Rat & Mouse Club, the American Fancy Rat & Mouse Association, and the Rat & Mouse Club of America—hold shows, award ribbons and championships, and maintain breeding records not unlike those of the American Kennel Club. The web site of an English group, the London & Southern Counties Mouse & Rat Club, shows pictures of more than forty standardized varieties of fancy mice, which come in five categories: selfs, which are the same color all over; tans, with tan bellies no matter what the color of the rest; marked, with spots in patterns or random spots; satins with silky sleek coats; and AOVs. The abbreviation stands for "All Other Varieties" and includes some mice that are fancy indeed. One variety has long, silky hair, and another, the Astrex, sports curls, even in its whiskers. And here are some—only some, mind you—of the colors assigned to fancy mice: champagne, chocolate, cream, dove, fawn lilac, red, silver, and white. I find myself grateful that Hannah, who would have needed my help, did not compete in fancy-mouse shows, for that would have meant far more transport than angelfish ever required.

Unlike the fancy mouse, *M. m. laboratorius*, the laboratory mouse, is not one that people mean to fall in love with. And if by chance they do, woe betide them. Sarah's story, soon to be told, will illustrate that point. Suffice it to say right here that ninety-five percent of all animals used in biomedical research are rodents, and of that number, fully ninety percent are rats and mice. The other rodents used are mainly hamsters and guinea pigs. Laboratory mice have been bred to create tens on tens of strains that carry heritable mutations in physical, physiological, and behavioral characteristics. Thus, we have genetically homogenous strains of diabetic mice, blind mice, hairless mice, mice with Lou Gehrig's disease, mice prone to various cancers, alcoholic and drug-addicted mice, pathologically obese mice, and a wonderment of others, all employed in studies related to the characteristics of a specific strain. (It should be noted that these mice, however peculiar in appearance or behavior, remain mice, able to revert to their original wildness and to breed with their country cousins.) Since 1980, the ways in which many strains are now put together results in a creature that some people would consider—unfairly, I think—a Frankenmouse. In the early days of strain-development, Mendelian methods were used to select and breed mice with particular traits. Nowadays, scientists employ genetic engineering. A paper from the National Human Genome Research Institute gives this description of making mice to order:

> One of the most important advances has been the ability to create transgenic mice, in which a new gene is inserted into the animal's germline. Even more powerful approaches have permitted the development of tools to "knock out" genes, which involves replacing genes with altered versions; or "knock in" genes, which involves altering a mouse gene in its natural location.

Transgenic strains are valuable indeed, but some of them show an unmousely failure to reproduce well. To counter low birthrates, researchers use cutting-edge technologies such as *in vitro* fertilization and ovary transplants.

The mapping of the house mouse's genome was completed in 2000, and for a good reason: because it possesses a version of almost every human gene, *M. musculus* can provide researchers with means for investigating the genetic bases of human disease as a prelude to discovering treatments. As the Genome News Network puts it: "The mouse genome is essentially a reference manual for understanding the human genome. Virtually every gene in the mouse is also present in people, and the neighborhoods in which these genes reside are strikingly similar in humans and mice." And mice are certainly much smaller, easier, and less ornery to work with than people. The genes of both mice and men arose in those earliest true mammals, the Pantotheria, millions upon millions of years ago in the Middle Jurassic. What, then, makes Us different from Them? The simple answer can be summed up in a single word: evolution. The Genome News Network adds a little detail: "Though many human and mouse genes appear to be similar, they may have taken on slightly different roles, or be active at different times during the life of a person or a mouse. How genes are controlled may help determine whether we have hair or fur or other distinct characteristics."

By some sort of miraculous coincidence, I know one person who worked for a recent year with laboratory mice and a second who is currently possessed by them and shall be so for some time to come. The first is Jay Hirsh, molecular biologist and professor at the University of Virginia. His research subjects have, for the most part, been fruit flies, not mice. The neural pathways of fruit flies are remarkably similar to those found in vertebrates, including, of course, mice and people. And there's a maxim to be found in this phenomenon: when a biological arrangement is successful, evolution does not need to tamper with the basic model. Not only developmental but neural and behavioral genetics may be examined in fruit flies. Jay's particular work with the little flies has involved subjecting them to crack cocaine and observing the progressive stages of intoxication, which are, not incidentally, the same stages found in rodents and humankind. The point of these studies is that pathways found in these tiny insects may lead to understanding of the pathways in vertebrates—in this case, the route by which addiction occurs. When the chance came to spend a year working with mice to the same ends, Jay took it. The move produced results other than those hoped for. Jay writes wryly that he and his students "ended up showing that mice were smarter than us and, also, that it's really boring being housed in a mouse cage. We tried adapting some fly technology to self-administration in mice, but they were doing the behavior not for the rewarding property of cocaine but from boredom." Fruit flies are again ascendant in his lab.

But Jay's daughter Sarah Hirsh is dedicated to the mice. A graduate student in neurobiology in the biology department at the Johns Hopkins University in Baltimore, she is part of a group studying various aspects of the sense of smell in *M. musculus*. (Another study in the biology department uses mice to investigate nerve growth, and yet a third study looks into how light regulates the circadian clock.) While some members of Sarah's lab focus on the function and physiology of the cells responsible for detecting the smells that reach the mouse from elsewhere, she concentrates on the development and regeneration of cells within the olfactory system. And she gives me an amazing fact: the olfactory epithelium, which is the mucus-covered tissue that lines the nasal cavity where smells bind to receptors, regenerates continuously throughout an animal's lifetime if it is subjected to physiological stress or damage. Sarah's specific project is to examine the regenerating subset of cells within the epithelium. "In mice," she says, "we can perform manipulations in a living animal by using cell-specific proteins to label or remove subsets of cells. We can also take tissues and analyze them outside of the mouse." She dissects the epithelium, separates out the cells capable of regeneration, and sets up primary cultures. That way, she can identify individual cells and study the effects of various factors on their growth and survival.

Why use mice for examining the workings of the nose? Sarah says, "Mice are a good way to study the olfactory system. The morphology is similar to that of humans, with some components possessing a greater variety of amplitude because of the increased importance of smell for mice." People have some ten million receptor cells, a number that seems high, but mice—and cats and dogs—have untold millions more than that. Sarah adds, "The olfactory system is a useful tool for studying general neurogenesis and regeneration." In other words, she looks at how nerve cells are created in the nose and how they may later recreate themselves, and she examines the olfactory bulb, located in the front part of the mouse's brain. It looks something like an elongated daffodil bud on a slender stem. The bulb performs several jobs after it receives input through a web of specialized nerves from the olfactory epithelium. Among other activities, it enables the mouse to distinguish among various smells, separating, say, edible cheese or a choice cricket from something unpalatable; it also acts as a filter that lets the mouse home in on a few smells, like those of food or sexual pheromones, rather than on the whole A-through-Z array that is present at any given time. Our own olfactory bulbs, which are located beneath the cerebral cortex, not in the front, work in the same way, but if human survival and the continuation of our kind depended as fully as a mouse's on the sense of smell to find food and mates, rather than on our other senses, we'd also have many more odorant receptor proteins and our olfactory bulbs would be far larger. Imagine how unceasingly twitchy our noses would be! The marvel is that epithelial cells in the human nose can regenerate as well as do those of mice. Study of Them is more than repaid by the knowledge gained about Us.

Trying to see Sarah's mice more clearly, I ask if the animals that she works with come in various colors or just one. She responds that different strains are different colors. "I have black, brown, and white mice," she says. "Sometimes, people cross or mix cells from two different colored mice in order to track the percentage of the genome that came from one parent versus the other." I ask, as well, how she views her mice. Does she find them cute, beautiful, plain, stupid, smart? Are they likeable or not? Her reply is worth setting forth in full:

"They are cute as they age. At birth, they are pink nuggets about the size of a pinky toe. They start to grow fur at about seven or eight days and open their eyes close to two weeks old. When I give them ear tags at three weeks of age, they are very active in the 'popcorn' stage, jumping all over the place, and still very cute. They're pretty smart, able to tell when I've been manipulating their brothers and sisters. They know they should avoid me.

"But I learned my lesson on getting attached to mice. The mice are usually housed downstairs; so we don't see them for most of the day unless we go down to work with them. A couple of weeks ago, I ended up inheriting a cage with a mother and a one-day-old litter in the lab. (Once cages come upstairs, they are considered 'dirty' and can't reenter the 'clean' facility downstairs.) I watched the mother raise her young and found her to be very entertaining—she'd nurse for a while, get tired, and retreat to the other side of the cage looking exhausted. The pups were really cute as well, and I got to see them open their eyes for the first time while I was changing the cage. When they were about three weeks old, I took a couple to use in my studies but didn't need the mom or the rest of the litter. I didn't know if I could get rid of them—I had become attached at this point—and kept them on my lab bench for another few days. Finally, I asked someone else to sacrifice them for me; it would have been too hard to do myself. That's the last time I'll let myself watch baby mice grow up."

A painful lesson, but Sarah and her colleagues in all sorts of mouse studies know that the deaths of laboratory animals are necessary in order to perform their work. One researcher concerned with keeping mice as comfortable as possible while they are in service to humankind has written, "Since we owe them a great debt for their involuntary contribution to science and medicine, it would be proper for us to ensure that their quality of life is as good as we can reasonably make it. This is made somewhat easier because laboratory mice are not very interested in people, but are passionately interested (for good or ill) in other mice." Some suggestions for ensuring the comfort of lab mice are keeping social groups small—no more than five or six in one cage, providing cages with solid floors and plenty of room to run, and maintaining low levels of light for these primarily nocturnal animals.

What has the house mouse contributed to science and medicine? To give all too few examples, the house mouse has figured in studies of memory storage, the pathogenesis of muscular dystrophy, the potential toxicity of artificial sweeteners, and effects of light on the development of circadian rhythms. Circadian rhythms are the daily activity cycles of just about every organ, from brain and heart to liver and lungs; they govern periods of rest and wakefulness. One 2006 study investigating the relationship of light to these rhythms reported that the brains of baby mice exposed to light twenty-four hours a day were prevented from setting the biological clock into its usual benign cycles and that such disruption would have a lifelong impact on behavior. These findings soon made the leap from baby mice to baby human beings. It was customary—and still is in some neonatal intensive care units—to keep premature babies under the lights around the clock. But when night/day cycling is adopted, the babies begin sleeping through the night earlier than those with constant exposure to light, and they also gain weight faster. The pain of mouse-sacrifice is surely assuaged by joy in improving the well-being of babies who refuse to stay womb-bound until gestation's normal end.

Another study well worth mentioning is one that explored the relationship of exercise to mental functioning. Researchers housed some mice in cages furnished only with food and water, while others were given the benefit of a turning wheel. Mice are maniacs when it comes to running; those with wheels clocked an amazing average of five kilometers a night. After six weeks the two groups were tested on their ability to make their way to a difficult objective. The couch potatoes were significantly slower to learn new tricks than were the runners. The latter had grown more new neurons in the hippocampus, which is the part of the brain with jurisdiction over learning and memory. It has been demonstrated that exercise can also trigger the growth of new brain cells in adult human beings. There's an almost Aesopian moral to the story: In mice and men, the more muscles exercise, the better the brain. (As an aside, the word "muscle" comes from *mus* and translates directly as "little mouse." Two theories explain the relationship: one, that the rippling movement of a muscle resembles that of the little animal, and two, that a muscle has the shape of a crouching mouse, rounded with its rump up.)

*

With Aesop, we are led to *M. m. litterarius* and its most variable habitats, those in fiction, fable, and verse. And the literary mouse comes in many guises. It appears to children in nursery rhyme and song. The round "Three Blind Mice" has been sung since the early 1600s. The very existence of the trio has been questioned by poet laureate emeritus Billy Collins, who asks, among other things, "Would it not be difficult for a blind mouse / to locate even one fellow mouse with vision / let alone two other blind ones?" But we know the answer to that: with its extremely sensitive olfactory system, no. In one of the most modest appearances of the literary mouse, he (or she) is as anonymous as the three blind mice—the little mouse that Pussy Cat frightened under a chair when he came to London to visit the queen. I am, however, willing to give him (or her) a name, for the small animal is none other than *M. musculus*, the house mouse.

It is in Aesop's fables that the mouse often takes center stage, sometimes happily helping larger animals and sometimes sadly becoming their dinner. Most of us know of the mouse that gnawed the ropes with which the hunter bound the lion and so set him free: Even a meek mouse can help a regal lion. And we have met the country mouse who accepted the invitation of the town mouse—"Come visit me! The world offers better eats than rural beans!"—only to have the fine meal interrupted time and again by large, loud people or furiously barking mastiffs: Better beans in peace than cakes and ale in fear.

The tale of the mice and the weasels may be less familiar. The two tribes were ever engaged in bloody war, and always the weasels won. In council, the mice decided that the reason for their continuing defeat was that they had no chain of command. So, they chose as generals mice well known for their family lineage, courage, strength, and ability to give good advice. The newly appointed generals proceeded to assign their troops to companies, regiments, and battalions. Then, they sent a herald to proclaim war against the weasels. So that all the mice could recognize their commanders, the generals put on helmets of straw. But no sooner had battle begun, than the weasels gained the advantage and routed the mice. The troops fled right smartly into their hidey-holes, but the generals, encumbered by their bulky helmets, could not follow them. The weasels caught and ate every last one. Moral: The more honor, the more danger.

The mouse inhabiting the poem "To a Mouse" by Robert Burns (1759–1796) is one whose nest he turned up with his plow in November 1785. She is a "wee, sleekit, cowrin, tim'rous beastie," to whom he apologizes for frightening her and ruining her tiny house.

I'm truly sorry Man's dominion
Has broken Nature's social union,
An' justifies the ill opinion
Which makes thee startle
At me, thy poor, earth-born companion,
An' *fellow-mortal!*

He sighs and shrugs. "The best-laid schemes o' *Mice and Men* / Gang aft a-gley." But he pronounces her more blessed than he, for she lives solely in the present moment, while he must contemplate an unforeseeable, unknowable future that he can only guess at and fear.

Most likely, Burns would have been an inept researcher, incapable of sacrifice, but he has a keen sense of mousekind's honorable place in the scheme o' things. And he rightly laments the effects of human dominion on the natural world.

Mice have found an abundance of suitable literary habitats in the twentieth century. Burns's wee beastie and her cohorts have moved into both the title and the text of John Steinbeck's 1937 novel *Of Mice and Men*. Its protagonists, George and his big, mildly retarded companion, Lennie, undertake jobs on a ranch. Not only do their plans—their very lives—gang a-gley but also the lives of a mouse and other soft creatures, which Lennie, caressing them with his huge, strong hands, literally loves to death. Stuart Little stars in the eponymous tale by E. B. White, which appeared in 1945; the second son of a New York family, he is not only very small but has the features and the modest manners of the best of mice. And he is brave, going off on a perilous journey to find his friend Margalo, the bird saved by the Little family. We meet the mouse Reepicheep in three volumes of C. S. Lewis's *The Chronicles of Narnia*, which was published not long after *Stuart Little*; wielding a rapier, Reepicheep is invariably brave in battle (but not so stupid as to wear a straw helmet). When he loses his tail, the emblem of his honor, he and the mice under his command finally persuade the lion, Aslan, to restore it. Mice also star and struggle in Art Speigelman's graphic novels *Maus: A Survivor's Tale* (1986) and *Maus: And Here My Troubles Began* (1991), which deal with the Holocaust and its reverberations. Spiegelman portrays the Jews as mice, meek and helpless victims of the Nazi cats, who toy with their prey before killing them. But, as with all mousekind in more than a hundred million years of history and prehistory, these mice have staying power. Many survive physically, and the others live on in remembrance.

What of Mickey Mouse? He is by no means *M. musculus* but belongs to a genus comprised solely of himself and Minnie: *Murinoides disneyi*, or "Disney's mouselike creature." Kept forever youthful by his enormous eyes, he is a showpiece, an entertainment—diverting, yes, but hardly to be taken seriously.

*

We do take real mice seriously, not just in the lab but at home when the commensal relationship disintegrates into Us vs. Them, and we become, willy-nilly, predators. And we have devised weapons a-plenty. Those of us with tender hearts may use live traps. If we do, we need to take the mice miles away lest their olfactory systems lead them rapidly right back home. For the rest of us, ingenuity has created a bogglement of killing variations on the conventional snap trap. One is a zapper that administers a lethal electric shock to the mouse or rat that enters it. Others lure mice into a baited environment from which they cannot find egress. Yet others feature glue from which mice cannot un-stick their feet; one of the selling points here is that the glued mouse may be bagged and tossed while still alive, thus eliminating the odors of decomposition. A homemade method of dispatching mice consists of filling a five-gallon bucket with water and building a ramp that leads to the top; the curious climbers tumble into the water and drown. Rodenticides also offer a means of doing them in. Some act as anticoagulants, some damage the central nervous system, and some cause death by other means. It is not easy, however, to poison a mouse for the animal nibbles, and if the first small bite does not bring on malaise, it takes another bite. And, as Jay Hirsh says, mice can be smarter than we are.

Or possessed by a different agenda. A friend, who lives with her husband at the end of a mile-long gravel road out in the county, noticed that the cabin on their property had suffered serious infestation by mice, which had used the cabin night after night as a dance hall and left their calling cards everywhere. The place was not fit for putting up guests. So she and her husband laced the cabin with green rodenticidal pellets. Behold, the pellets disappeared, but, to their bewilderment, the signs of mouse-dancing did not. It came to pass that one day, her husband wanted to start an old pickup truck that had languished immobile for some time. But nothing could coax it to start, not even imprecations. When he looked under the hood, the first thing that he did was summon his wife. "The innards of that truck," she told me, "had served as mouse-housing for a good some time. And, looky there, the walls of their nests were inset with lots of little green accents. Did you know that mice are interior decorators?"

*

Mice have been with us forever and may conceivably outlast us. Their reproductive strategy is similar to that of the dandelion or the frog: strength in numbers. They produce litter on litter of little pink pups in order to maximize the chances of survival for the species, if not for the individuals.

Melanodon would be right proud.

TEN

Red in Every Season

{ A Tale of Shade, Sunlight, and Storm }

On Valentine's Day, at six in the morning, just before night has begun to dissolve into day, the crash occurs. It wakes me. The sound is that of a heavy accumulation of ice sliding off a neighbor's roof and shattering. The supposition is not amiss, for a heavy fall of sleet and freezing rain started just before dark on the previous evening. During the night, the world froze solid. In the morning sunlight, it sparkles. Icicles hang from branches and power lines. The boughs of my Norway spruce, usually lifted upward in a cheer, are weighted down and cheerless, and my small corkscrew willow is bent nearly double. But it is not till I come downstairs and look out the kitchen windows into the backyard that I see the reason for the crash: the storm has caused the old red maple in the yard next door to lose its hold on life. Thunderously uprooted, it lies across the entire backyard of my young neighbors and a good half of mine.

My first thought is that hundreds of pesky red maple seedlings will no longer be popping up in my vegetable gardens, at least not after last year's keys have had their chance to sprout. A corollary is that, without the arboreal screen of that big tree, my gardens will receive more light. The second thought is that my chain-link fence is in sore need of repair. The tree's downward impetus has bashed a V into part of the fence's metal railing and dislodged the railing elsewhere. And the third is sheer, smiling delight at watching the gray squirrels and a myriad birds—cardinals, blue jays, juncos, chickadees, titmice, song and house sparrows, house finches, goldfinches, and more—romping in the debris. It is as if the branches form an intricate jungle gym. If those branches did not cover much of the area used for tomatoes, squash, and scarlet runner beans, I'd let them stay so that the winged and four-footed creatures could have ready access to a playground that they clearly enjoy. Only after my mind has entertained these notions, do I remember to feel grateful that human habitations have not suffered damage, nor have we, the inhabitants.

Four days later, there is no hesitation about feeling grateful when Adam, the young man who lives next door, brings his borrowed chainsaw into my yard and begins the cleanup. All that's left for me to do is gather twiglets and put them in the barrel that I use for garden waste. And the twiglets tell a tale. Till February, the weather was unseasonably mild; January gave us temperatures in the low seventies. Two weeks early, the maple began to respond to springlike warmth by setting copious clusters of buds. True to the tree's name, the buds are red, a sweet soft red. Had they bloomed, the female flowers would have been crimson, far more flamboyant than the male flowers, which are hardly more than pale pollen-bearing stamens poking out of the twigs. The paired keys, light green with a reddish flush, would have started helicoptering earthward on their sturdy wings in late May or early June—and would have sprouted the following year in my garden, well before the butternut vines had even begun to sprawl beyond their hills or the pole beans to climb their tower. The newly unfurled leaves would have been red, too; in spring it's easy to spot red maples, for they abjure spring green for a color that serves to protect new foliage from sunburn. The mature leaves would also have borne the tree's signature red on stem and midrib and, in the fall, they would have blazed with a red deeper and more purple than that of orangey-red sugar maple leaves.

The tree was one of a very large family, the Aceraceae or Maple family. Its botanical name, *Acer rubrum*, translates directly as "red maple." The Latin word *acer* actually means "sharp." That's an apt description of the leaves' pointed lobes. But it actually comes from the metal point on the spear handles that the Romans shaped from the maple's hard wood, and the word slid from spear-point to the whole tree, roots, trunk, and crown. The ancestral maple appeared on this earth with the other flowering plants about 120 million years ago in the Early Cretaceous Period. Since then, some 125 species of maple trees have grown in North America, Europe, Asia, and North Africa. Some are small, like the elegant Japanese maple (*A. palmatum dissectum*, "deeply cut palmate maple") that spreads its slender branches in my front yard. Boxelder (*A. negundo*, roughly "resembling the chaste-tree"), another tree usually on the small side, is one that noted naturalists Donald Culross Peattie and Elliot Coues have respectively called "the Maple's poor relation" and "this small bastard maple." Why take such a patronizing tone? Because the short-lived boxelder has soft, weak, well-nigh useless wood and, worse, has made a weedy nuisance of itself, for it spreads as readily as sumac and ailanthus. Others maples are huge, like the native red and sugar maples (*A. saccharum*, "sweet juice") and the Norway maple (*A. platanoides*, "resembling a plane tree"), which was introduced into the U.S. to shade yards and dignify tree lawns. In the early 1950s, my father planted a Norway maple in his backyard; by the time that he died in 1972, it was as tall as his three-story house. His ashes now rest at its base.

Looking across my crippled fence, I see the great chain-sawed rounds of the red maple's trunk. Those from the base to what would have been the five-foot mark are more than two feet in diameter. A few show slight hollowings from rot, but most seem sound. Adam carted off a truckload of twigs and small branches several days ago. Now he splits the rounds into firewood. There is some good after all in the storm-toppled giant.

The sugar maple comes close to being everybody's darling. Festivals mark sugaring off, the distilling of the sap into syrup and sugar. I think of Monterey, Virginia's Maple Festival, an event that brings in thousands of tourists on the second and third weekends in March, and has done so for fifty years. What an old-fashioned whoop-de-doo it is—the Maple Queen contest, the Maple Queen Ball, the Festival Fling Ball, blue grass and gospel music, clogging, craft shows, and all manner of home-cooked food, especially plain or buckwheat pancakes drenched with syrup. *A. saccharum* puts the other members of its genus in the shade. Its sap has been used for sweetening purposes since prehistoric times. John Lawson, appointed Gentleman Surveyor General of North Carolina, encountered this tree on his travels. Not sure that it really was a member of the maple tribe, he nonetheless gave the opinion in 1709 that it "may be rank'd among that kind." And :

> The *Indians* tap it, and make Gourds to receive the Liquor, which operation is done at distinct and proper times, when it best yields its Juice, of which, when the *Indians* have gotten enough, they carry it home, and boil it to a just Consistence of Sugar, which grains of itself, and serves for the same Uses as Sugar does.

In his catalogue of the trees of the Carolinas, he describes most trees, from the oaks to the pawpaw, with minute detail. But, aside from the description of the sugaring-off process connected with the maybe-maple, he gives only one short sentence to the other maples: "The Maple, of which we have two sorts, is used to make Trenchers, Spinning-wheels, *&c* withal."

Though my wager will never be called, I'm willing to bet that one of the two sorts is the red maple. I'll also bet that John Lawson didn't know that all maples—red maple, silver maple, black maple, and even the bastard boxelder—yield sap that can be boiled down into syrup and granules of sugar. The sugar maple's advantage is that its sap has a higher sucrose content than does that of its congeners. Other early naturalists, praise be, paid far greater attention to maples in general, and to the red maple in particular. One was the Englishman Mark Catesby (1683–1749), who explored the Atlantic seaboard from Virginia through the Carolinas into Florida. Along the way, he collected a prodigious number of botanical specimens and made drawings of the unfamiliar plants, birds, and animals that he encountered. On his final return to England in 1726, he began assembling his life's work, several volumes devoted to the natural wonders of the New World. Behold: a yellow-throated creeper perched on a twig of the red-flowering maple! The bird is known today as the yellow-throated warbler; the tree is firmly in the ancestral line of the once-towering red maple that grew next door. Catesby took many plants, including poison ivy, back to England, and he reported that red maples "endure our English climate as well as they do their native one, as appears by many large ones in the garden of Mr. Bacon at Hoxton." Hoxton, then a suburb just north of London, no longer boasts such a garden, but since Catesby's time, the red maple has certainly settled itself handsomely into the English landscape.

In 1748, Peter Kalm (1716–1779), a Swedish botanist of Finnish descent, embarked on a North American plant-gathering expedition that lasted for the next four years. He had been appointed by the Swedish Academy in the person of Linnaeus, under whom he had been a stellar student, to gather seeds and specimens that might be useful to Swedish agriculture—tough plants that could survive the rigors of Swedish winters. Almost immediately on arrival, Kalm was befriended by none other than Benjamin Franklin and the botanist John Bartram. His extensive travels, ranging from New England and Canada to northern Virginia, resulted in a minutely detailed journal on everything that tickled his curiosity: asparagus growing wild in a sandy soil in New Jersey; a May hatch of locusts that chirred in an almost deafening fashion; the onrush of a late-spring tornado; the New York and Boston prices of staples like flour and rum in 1720; the 1749 prices paid for wild animal skins in Canada; the rheumatism and pleurisy

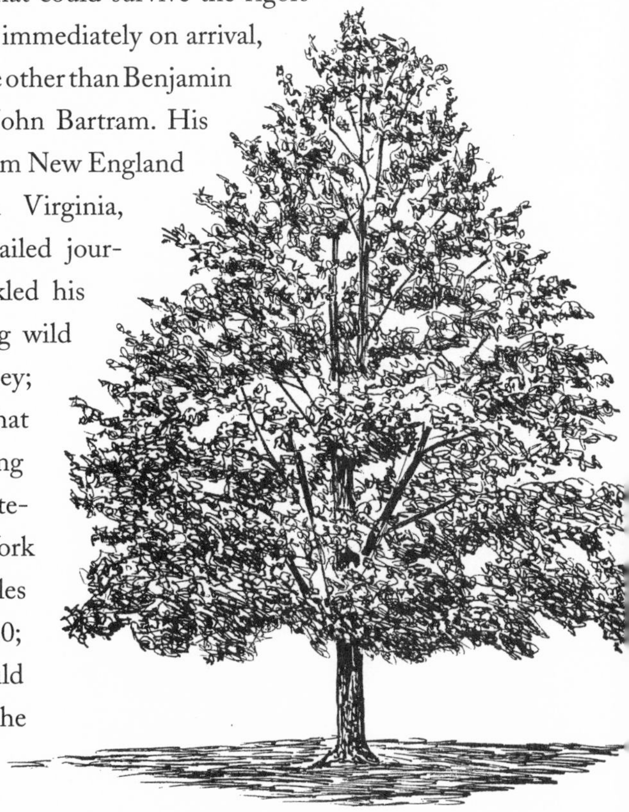

that afflicted Canadian Indians; an October foray to collect the seeds of sugar maple, hornbeam, linden, and ash to take back to Sweden; and a superabundance of other topics, including the first-ever eyewitness description of Niagara Falls. Lobsters, parsnips, raccoons, rattlesnakes, fireflies, coal—little escaped his diligence. Nor did the red maple fail to attract his keen eye. In the entry for October 6, 1748, he noted that the tree was plentiful near Philadelphia, especially in swampy places, and he proceeded to list the manifold uses to which people put it. Kalm was well apprised, too, of red maple's sweet secret:

> Out of its wood they make plates, spinning wheels, spools, feet for chairs and beds, and many other kinds of turnery. With the bark they dye both worsted and linen, giving it a dark blue color. This bark likewise yields a good black ink. When the tree is felled early in spring a sweet juice runs out of it, like that which comes out of our birches. This sap is not made use of here, but in Canada they make both syrup and sugar of it.

On Kalm's return to Sweden, Linnaeus credited him with bringing home ninety new species, sixty of them hitherto unknown in the Old World. He is honored to this day by the genus name that Linnaeus bestowed upon the mountain laurel, *Kalmia latifolia*, "Kalm's broad-leaved plant," and the sheep laurel, *K. angustifolia*, "Kalm's narrow-leaved plant." His name also figures in the species designation of one St. John's wort, *Hypericum kalmianum*.

Historical notice of *A. rubrum* hardly stops with Kalm. Giving swamp maple as the tree's common name, John James Audubon (1785–1851) gives it pictorial laud in his painting of red-winged blackbirds. One coal-black male bird flies, showing his jaunty red epaulets trimmed with gold. Another male, a boldly striped brown female, and a juvenile bird perch on still leafless twigs that have just burst open into red-pink bloom.

Catesby and Kalm—I see them in mind's eye as they traipse through the American countryside. Their traveling kits held no plastic, no aluminum, no Gore-Tex. Natural zeal led them even into the wilderness dressed like gentlemen in waistcoats, breeches, stockings, and hats. Other people, similarly encumbered in the way of clothing, accompanied them to tote the collecting cases, writing desks and inks, and sundry other provisions. Hardship, of an overburdened kind that I can barely imagine, was evidently an accepted concomitant of extending and enriching scientific knowledge. They and others like them—John Lawson and Audubon, for two—accomplished wonders with the relatively primitive technology of their times (though Audubon was given more to comfortable shirtsleeves than to fancy dress). They were as brave—and just as stubbornly determined—as were Columbus and the first men who walked on the moon.

My young neighbors need a tree. They will miss the shade that the giant provided from the intense sunlight of a summer afternoon. I hope to give them one of its children. For, by an immutable act of nature, more than one shall sprout amid my vegetables.

ELEVEN

Big Beast

{*A Story of War*}

"Saw Big Bill in your yard this morning," says my neighbor Joe. The very next day, I, too, see Big Bill, who may, for all I know, really be Big Betty. Unlike many other animals that wear their apparatus in full sight, there's no eyeballing the gender of a groundhog. Easier to call this chubby, buck-toothed, grizzled-brown creature Big Beast. It is sitting upright on its haunches, forepaws holding a sunflower to its mouth.

And the jaws are working rhythmically. I can spare a sunflower, though, for the birds planted a golden multitude of them below the feeder that's stationed in the backyard vegetable patch. The next minute, when I go outside, Big Beast right smartly assumes an all-fours position and scampers away. In almost no time, I discover that it has consumed not just one sunflower but nipped off the top leaves and buds of at least ten, nor are sunflowers the only items on the menu. Yesterday, the sprouts of the scarlet runner beans had presented their initial set of heart-shaped secondary leaves. Today, nothing is left but denuded inch-high stems. The same is true for half of the Kentucky Wonder pole beans. And it's only mid-May. The prognosis for the rest of the summer is not good.

This is not the first year in which the backyard garden has been raided by Big Beast and, I am sure, its parents, siblings, uncles, aunts, and cousins. The garden was created in 2002. That summer, tomatoes, green beans, and cucumbers were set into the good brown earth. The tomatoes and beans yielded amply (in 2002, beans were evidently not at the top of the chow-down list), but every last one of the cukes fell victim to raging herbivory. Since then, cucumbers have been grown with joyful abundance in the front yard, which offers Big Beast no hidey-holes.

Unilaterally, I declare war. But just who is my enemy?

Marmota monax, that's who. One enthusiastic web site translates the binomial as "marvelous mammal," but the name really means "solitary marmot," for this rodent is the only marmot that prefers its own company to that of others—except when a male seeks a mate or a female rears her cubs to self-sufficiency. The word "marmot," which made its way into the English language in the 1600s, comes from the French *marmotte,* which in turn may spring from a Romance term *murem montis.* That last means "mountain mouse," a nod to the fact that *M. monax*'s European compeers, of the genus *Arctomys,* "bear mouse," live in the Alps and the Pyrenees. The gourmand in my garden, a native of North America, is found in most of the states east of the Mississippi, as well as in Alaska and much of southern Canada. It shares its genus with seven Old World and six New World species. Members of the Sciuridae, the Squirrel family, all are basically enormous, chunky, tailless ground squirrels. Like all squirrels, they are rodents, kin to mice and rats. Like all rodents, they possess chisel-sharp incisors that don't stop growing but must kept in trim by continual gnawing, be it on nuts or sunflowers. But unlike most rodents, groundhogs behave like bears come cold weather and go into true hibernation, sinking into torpor in late October. Depending where they are, they rise again anywhere from late February to early April. As the length of daylight triggers birds' migrations, so air temperature determines when Big Beast wakes.

But Punxsutawney Phil emerges yearly from his burrow on February 2, Groundhog Day, with his forecast of winter's duration: If he sees his shadow, winter will continue to hold earth in thrall for another six weeks, but if the sky is clouded over so that he casts no shadow, then spring is already hastening to turn the land green again. This peculiar form of weather prediction began on February 2, 1886, in the eponymous town of Punxsutawney, Pennsylvania, which lies about sixty-five miles northeast of Pittsburgh. Until 1966, eighty years later, the ceremony was held in secret and only afterwards was Phil's prognostication publicly announced. Since then, the day has become a major whoop-de-doo, luring multitudes of groundhog fans and tourists, not to mention the media, to this town with a current population of slightly more than 6,000. The event has also led to the 1993 making of a movie, *Groundhog Day*, starring Bill Murray as an egocentric TV weatherman who visits Punxsutawney on Groundhog Day only to find that the occasion repeats itself without cease until, at last, he manages to scrape the calluses off his long-hidden human kindness.

Why choose February 2, a day on which the sleep of a normal groundhog in Punxsutawney or, indeed, anywhere else can't be broken? The Germans who settled in that part of Pennsylvania brought with them a tradition associated with Candlemas Day, which occurs on February 2. I find a translation of a German verse that puts it this way:

For as the sun shines on Candlemas Day,
So far will the snow swirl until May.
For as the snow blows on Candlemas Day,
So far will the sun shine before May.

In Germany, the animal associated with shadow-casting was the badger. One species of badger does live in North America, but it's confined to the central and western United States and central Canada. Its formal name is *Taxidea taxus*, which bit of taxonomy-speak translates as "badger resembling a badger," and it's a member of the Mustelid family, the one to which otters and skunks belong. So, badgers being in short supply, the groundhog, an animal easily found in Pennsylvania, was brought in to play the part. And the play is elaborate. These days, accompanied by thousands of followers, twenty-one gentlemen dressed in top hats and tuxedos with black bow ties make an early morning trek to Gobblers Knob, a wooded hill not far from the town. The gentlemen comprise the Groundhog Club Inner Circle, and each bears a title proclaiming his function, including Burrow Master, Cold Weatherman, Fair Weatherman, His Scribe, His Handler, and Stump Master. The Fair Weatherman, an attorney by trade, is also Phil's lawyer, guarding the animal's trademark and handling licensing deals. The role of Stump Master is particularly important, for Phil, who is on display at all other times in his special, climate-controlled den at the Punxsutawney Public Library, must be transported by this dignitary to an artificially heated stump atop Gobblers Knob. Subsequently, at the magical hour of 7:25 A.M., he is lifted therefrom by His Handler. Aha! Phil's unseasonable wakefulness is triggered by a temperature that tells him that Spring has come and that it's time to seek a mate. But all he gets is a trip back to his cold den in the library. The point is not making a forecast but rather creating an occasion for a bunch of guys to dress up and have a great good time. For that they can't be castigated. But I wonder if any of them have ever been victimized by a ravening groundhog that seeks sustenance from *their* sunflowers, *their* green beans, *their* squash.

Groundhog is a fit name for this particular squirrel, which digs commodious burrows with its long, curved, tough claws. Literally, it lives in its digs. Construction of a single burrow can call for the excavation of 700 pounds of soil. It has at least two entrances, sometimes as many as five and may consist of more than fifty feet of tunnels, plus various chambers for hibernating and nesting. The burrow system is kept scrupulously clean, with wastes taken outside and buried. Big Beast bears several other common names. One is whistle-pig for the shrill vocalizations made when the animal is alarmed. In fact, it employs a range of sounds from barks and squeals to grinding its teeth. I've heard one grunting—*huff, huff, huff*—as it climbed a tree to escape the playful leaps of my dog. A friend tells me that he's also seen the critter up a small tree, where it was munching happily on the leaves. A third name is woodchuck, which has nothing to do with its capabilities but is rather a mutation of the Cree word *wuchak*, which seems to have been a generic term for wild animals of a certain size.

How do groundhogs fit into the scheme of things? Their dining affects the plants that grow in their communities. In their abundance, they themselves provide food for other animals, like coyotes, snakes, and the larger raptors. Their old burrows provide homes for many other animals, such as rabbits and feral cats. They furnish sport for hunters and entertainment for undiscriminating lovers of wildlife. Some people like groundhogs exceedingly well, including the old man who lives two doors up the street. Early in the summer when the depredations in my garden had become noticeable, I heard him shouting to his middle-aged son that the groundhogs were *gone*. The shouting was occasioned by the fact that the old man is rather deaf. I shouted back, for the son to hear, that the groundhogs were not gone; they had merely moved two doors down. The son told me later that his dad considers the animals to be wonderful, God-given wild pets.

Certainly, they are not unattractive, with their large squirrel-like faces, small rounded ears, and short, bushy tails that they use for balance when sitting on their haunches.

Groundhogs have fans much farther afield than on my street. There are, of course, the top-hatted acolytes of the legendary Phil. And my daughter-in-law Debra reports admirers in France. She had an unexpected encounter with Big Beast during her stay in Paris in 1989, the bicentennial year of the French Revolution. She was there to research the art that Napoleon looted from various non-French sites as he marched across Europe, but all study and no play is not the way to spend time in Paris. So, Debra made a visit to the Jardin des Plantes, which is not only a botanical garden but also houses a zoo that opened in back in 1793. One section was devoted to North American mammals, including *le marmot amèricain*, the American marmot, which dwelled in an enclosure that Debra describes as "the centerpiece of the zoo. And what do you know, it holds a flippin' groundhog."

Recently, I spotted a vanity license plate reading WDCHUK in Manchester, New Hampshire. And at the grocery store just before Halloween, I came across a six-pack of a glass-bottled beverage called Woodchuck Draft Cider; pressed in Vermont, it contains five percent alcohol, and the groundhog on its label seems to be contemplating the deliciousness of orchard apples. On the other hand, groundhogs, gobbling up windfalls, can kill apple trees by destroying the roots with their tough claws. The creatures are serious agricultural pests, destroying crops and causing much damage to the livestock and farm machinery that fall into their vast burrows.

To get a farmer's point of view, I talk with Joel Salatin, proprietor of Polyface Farm, located in Swoope, Virginia, not far from the town in which I live. Joel is the quintessential organic farmer, raising chickens and cattle, maintaining truck gardens, catering strictly to local markets, and discarding nothing, not even chicken guts, which he composts and gives back to the good earth. When I ask him how he deals with the inevitable groundhogs, he quickly says. "We wage perpetual jihad on them. They're all right in the woods, but when they leave cover, we shoot them, we trap them, we chase them down and kick them. They are machinery-breakers and cow-breakers if one gets its leg caught in a hole." He adds, however, that he hasn't seen a new hole in his pastures during the past ten years—a few, very few, elsewhere on the farm, yes, but in his pastures, no. The reason for the absence of burrows, and of groundhogs therefore, is that he practices management-intensive grazing; indeed, he calls himself a grass farmer, for grass is a storage bin for solar energy and, so, the basic element in the food chain that sustains the rest of his enterprises. Grass grows in a two-week cycle, quiescent at first, then explosive, and finally quiescent again as the grass prepares to flower and set seed. Management-intensive grazing takes advantage of the explosive phase. So that the cows may dine on the most lush food, they are rotated frequently among Joel's pastures. How does such grazing affect the groundhog population? It reduces what Joel calls the "weed-load," and it keeps the clumpy grasses down. Close-cropped turf offers a groundhog little food and no protection from predators, like weasels, owls, and hawks that find groundhogs, especially the kitten-sized young, to be a fine source of fast food.

"How much wood would a woodchuck chuck if a woodchuck could chuck wood?" I survey the devastation in my garden and rephrase the words: "How much ground does a groundhog hog when a groundhog does hog ground?"

Sunflowers, as it turns out, have marvelous powers of recovery. In the axils of stem and savaged leaves, new buds form and burst into golden blossoms.

The problem is that Big Beast doesn't stop with sunflowers. The pole-bean sprouts—scarlet runners, Kentucky Wonders, Florida speckled butterbeans—are still very much on the *á la carte* list. Off to the hardware store I go to purchase two twenty-five foot rolls of two-foot-high chicken wire. Those rolls are so tightly wound that they do not wish to give up their coil, and it takes a peculiar dance to straighten them out. An end is put on the ground, about four feet of the roll are with great effort unwound, and, unwinding as I go, I stomp upon the edges. Once the stubborn wire has been flattened, it's cut into lengths that can encircle the bean towers. I dig a shallow trench around each tower, place the chicken wire in it, fold the edges together, and push the excavated dirt back in the trench.

Big Beast could easily dig under these new fences but does not. Instead, the creature chomps upon the butternut squash seedlings. Now I count on butternut squash not only for just plain eating but also for Christmas giving: it is the prime ingredient for butternut pies and butternut bread. Back to the hardware store I go for more chicken wire. But fences are not the only weapons in the arsenal. At the end of May, I leave a message in the voice-mail of the city's animal-control officer to see if it's possible to rent a live trap big enough for a groundhog. The officer sends me a five-page missive that includes three pages on the habits of groundhogs. As it happens, the city does provide traps but only for cats, not wild animals. And it issues warnings: "Cats are deemed personal property and may be the subject of larceny," and "Setting traps at night greatly increases the chance that you may catch a skunk and have to dispose of such animal."

Two thoughts occur to me: first, the city won't help me get rid of Big Beast and, second, I could easily get a trap by pretending that I want to catch a cat. It turns out that the animal-control officer will set and check a cat trap for three days; after that, an individual may rent the trap for a maximum of ten days at a cost of fifty dollars.

During the first week in June, I make an enheartening discovery: a live-trap large enough for a groundhog (or a raccoon, or a skunk, or—for that matter—a toy poodle) costs less than forty dollars, including tax, at a local builder's supply company. From a friend I hear of other ways to do in a groundhog. After one fleeing animal was stopped abruptly by a pea fence, she beat it to death with a two-by-four and flooded another out of its burrow, whereupon her husband shot it. A member of my exercise class suggests yet another, supposedly sure-fire method for banishing groundhogs; she heard it from a local pumpkin farmer, who was experiencing considerable ravaging of his crop: place two pieces of bubblegum, wrappers opened, into each hole, and in four or so days, the groundhogs will have up and gone. But a trap strikes me as the simplest, most practical, least energy-consuming solution. On the way home, trap in the back seat, I stop at the grocery store for bait: fresh apple, cantaloupe, banana, and peanut butter. By late afternoon, the trap has been installed near the scarlet-runner bean tower.

Six days later, it catches its first victim: a gray squirrel. Not what I'd hoped for, but I'm happy to learn that an animal as light as a squirrel can trip the trap's door. I release the squirrel into the yard. The only garden damage squirrels have wrought is picking immature eggplants not to eat but rather to play with. Eggplants, like cucumbers, now grow in the front yard.

Two days later, bingo! 3:15 P.M., and I'm about to leave the house to fulfill a volunteer commitment. A glance out the back window shows a critter in the trap—not nearly so large as Big Beast but a groundhog nonetheless, and it's in the station awaiting transport to Timbuktu. When I go outside to make a quick investigation, the animal does not move. Only its sides rise and fall almost imperceptibly as it breathes. I've seen this same behavior in rabbits: if I don't move, then you don't see me. The trap is in the shade; Little Beast can wait till I return. It is not I, however, who takes the creature into exile. My reason for summoning help is pitiful: suppertime, and my body can't manage another hour before receiving sustenance. So, I call my brother, who arrives, puts animal and trap into the back of his SUV, and drives away. In the morning, the trap, scrubbed and shiny, is back in the yard. Of course, I set it again. No groundhog, no squirrel is lured into its innards; a young and unwary possum is the next victim. Like the squirrel, it is released into the yard.

The garden burgeons. The pole beans begin to climb, winding their way up the twine strung from tower bottom to tower top. The fenced-in squash put out fuzzy leaves the size of parasols. In my experience, once the plants gain mature leaves, the munching tends to stop. I remove the fences from the squash hills, and the vines merrily cover a patch of about 180 square feet. Buds appear, then yellow blossoms big as salad plates. The first flowers are male. In a week or so, the female flowers appear. The latter may be distinguished from the former by the miniature squash at the base of each flower. My experience, though, has been limited. On a tour of the garden one mid-August morning—checking tomatoes, picking the last of the carrots and the first of the Kentucky Wonders—I notice that the developing butternuts, full-sized but still green, have been savaged—the side eaten out of one, the base of another thoroughly gnawed. Those chisel incisors have been at work, scraping long furrows in the fruits' tender flesh. The trap, unset since the possum was caught, is again baited with cantaloupe. The bait disappears overnight, but nothing is caught. Several days later, droppings reveal the nature of the animal dining at the Havahart restaurant: mouse, and probably more than one, all of them too light to spring the trap. And the butternuts keep disappearing. What to do?

One of the five pages sent to me by the animal-control officer contains the names of people authorized to trap nuisance animals. Two specialize in beavers, another in bees. I call still another, Bryan Demory, who does not restrict himself to particular animals in his listing, and leave my name and phone number on his answering system. His own message is one of braggadocio: he is afraid of nothing, no, not even snakes. When he comes to my house, I greet a burly, amiable man in his early forties, who hands me his card: "Critter Git'r." Then he goes exploring. Half an hour later he reports groundhog holes in every yard up and down my street.

Mind you, the street lies only five blocks from downtown, not in the wilderness. There's even an old hole at the foot of a limestone drywall in my yard. And Bryan proceeds to give me a bit of instruction on groundhog behavior. When I tell him about the groundhog that I sent to Timbuktu, Bryan says that it is dead, for the animals are intensely dependent on their own territory. Taken from its accustomed digs, it is homeless, and it wastes away, unable to fend for itself.

So much for the merciful nature of live traps. Bryan also informs me that the animal does not travel far in its peregrinations, moving only 50 to 150 feet from its burrow during the early and late hours of the day. At night it sleeps. Only in spring, the mating season, will a male go farther afield. But this is August.

"He don't have to go very far here at all," Bryan says. "You've spread him a smorgasbord. And he won't dig a hole in your yard. There's a rule: You don't crap where you eat."

He asks then if I would be unhappy to have groundhog killed. His traps are dead-traps, square contraptions composed of slim metal bars. To set one, he has to use a tool that squeezes some of the bars into position for triggering. Even with the tool, setting the trap requires more strength than I can muster. He places the trap directly over the groundhog hole that seems most active. When the creature emerges, its head will touch the trigger and the rods will snap shut around its neck. Without hesitation, I tell him to go ahead, for the groundhog is killing my squash.

Later, it occurs to me that as far as the squash is concerned, I, too, am a predator.

Granted, I put the seed into starter pots and, a month later, transplanted the seedlings into hills. So, I have reason to think that the produce should be mine by rights. But what of the butternuts? Animals, including groundhogs and humankind, share an interest in avoiding suffering. It sounds ridiculous to bring plants into the equation, but many plants do mount defenses, from thorns to toxins, to keep themselves from being eaten. The strategy, however, is undertaken on behalf of reproduction: Let me live until the future of my line is assured. If the butternuts could speak, they would enjoin me to save and plant their seeds. I shall fail them. The large, flat ivory-colored seeds will be put into the trash, not the compost bin, where they would over-winter and later sprout when the compost is spread on the garden. No matter, for squash adheres to the-more-the-better rule when it comes to seeds, and the seeds are legion.

The very next evening, as dusk is coming on, Bryan marches into my backyard holding a very large animal that proves to be Big Betty. She has not been dead very long, for her body is still warm and completely without rigor. If I were my mother, I'd have taken the creature, skinned and dressed it, and braised it pot-roast style with carrots, onions, and potatoes. How on earth did my mother come to cook groundhog? In those days, when I was in my early twenties, married, and visiting several times a year, she and my father lived on a farm. He issued a decree: we will eat whatever this farm provides.

It provided kitchen-garden abundance, all the beef and pork that anyone could want, and a multitude of chickens, as well as apples and peaches. It also offered up snapping turtle complete with unlaid eggs, filet of timber rattler, and groundhog. My mother called it "woodchicken" so that my much younger sisters would not hesitate to eat it. But I think that they knew woodchicken's real nature all along. And on several visits I, too, partook. As might be expected of an animal that's a strict vegetarian, the dark meat was tender and delicious.

I look at the big furry beast suspended from the Critter Git'r's fist. A rush of sorrow takes me by surprise. An animal's life has been taken to save the lives of vegetables. When he says that he'll dispose of her, all I can do is nod. He puts her body under an old truck parked in the alley that runs behind my house. Her flesh will feed many appetites, from rodents to bacteria. It is some consolation to know that she will not be wasted.

TWELVE

The Mollusks in the Garden

{ *A Story of Earthbound Trails and Bungee-Cord Sex* }

Late spring through fall, morning sunlight reflects off the zigzag trails that cross my front walk. The snails that had been abroad to attend to their nocturnal gnawings on my plants are hunkered down now, hiding from heat, desiccation, and—though they cannot know it—being crushed by a heedless foot. Because I garden almost compulsively, I

want to know as much as possible about the denizens of my flower and vegetable beds. As always, I want to learn their names, on the premise that, even if I don't much like them and we shall never be friends, they cease to be complete strangers when I have their precise identities. They cannot know me (even though they do have a primitive brain), but that does not stop me from wanting to know them, and not just their names but their food preferences, their anatomy, their mating habits, and their very evolution. How did they arrive in my garden?

To begin with, the shells of the snails in my garden are beautiful, some completely the pale golden color of champagne except for a dark brown stripe no wider than a thread around the lip of the shell's aperture, and others of the same champagne gold but with broad chocolate-brown bands following the shell's whorls. They are right-handed snails, for the most of the shell is worn asymmetrically on the right side of the soft snail body. Left-handed species also exist. (There's a lovely word that describes the whorls and their position: chirality, which refers to snails' genetically programmed asymmetry.)

I think that I've properly identified these creatures, but I could be wrong; my opinion needs an educated confirmation or correction. I scoop up two banded specimens from the compost bin and, on the ground in the front yard, find several that are pure gold. Because it rained last night, the ground is damp enough for snails to venture out in the daytime. I put them all into a plastic bag, cover them with water, and place the bag in the freezer. The next day, after the snails have thawed, it is easy to extract the bodies from the shells with tweezers. Bearing newly cleaned shells, I drive to the Cooperative Extension Agency office in a nearby town; there, they are packaged for shipment to the experts at Virginia Tech. In two weeks, the answer arrives: banded wood snail, *Cepaea nemoralis*, a name that combines a Greek word for garden, *kephaios*, with a Latin word meaning "of the woods." The snail's name translates as "garden creature of the woods. It's sometimes called the grove snail, for *nemoralis* can also be translated as "of the grove." My guess at its ID turned out to be right. The banded wood snail has a genus-mate, *C. hortensis*, "garden creature of the garden," with the common name of white-lipped snail. The usual way of distinguishing one from the other is that the white-lipped garden species lacks the brown lip of its woods cousin. But an occasional morph of the garden species does develop a brown lip. The only recourse then is to examine its innards.

Neither of these snails is native to North America, though both are now widely distributed throughout the New World. Though banded wood snails may well have been introduced here as stowaways tucked snugly into imported plants, they were brought to the U.S. intentionally by Dr. W. G. Binney, a mollusk specialist, who wrote a tome entitled *Terrestrial Shells of the United States*. The beauty of the wood snail's shell apparently caught his fancy to the extent that he wanted to have the species in his very own yard. He told the tale of their arrival in America this way: "In 1857, I imported some hundred living specimens from near Sheffield, England, and freed them in my garden in Burlington, New Jersey. They have thriven well and increased with great rapidity, so that now the whole town is full of them." Now, 150 years later, *C. nemoralis* has settled into the Americas, North and South—nor has it done so at a snail's pace.

As the wood snails in my own garden amply demonstrate, they vary in appearance, and can do so dramatically. Nor, with or without bands, do they come only in a pale shade of champagne but may be bright yellow, pink, red, olive, or brown. They are protean creatures, polymorphs of many colors, and those with banding may be the luckier, for they are better camouflaged than their undecorated species-mates. Birds, in particular, find snails to be tasty morsels. Researchers report that the light-colored versions are most often found in monochromatic environments, like dunes and lawns, while the banded variety tends to show up in shadowy habitats like hedgerows and, I might add, the deep dark depths of my compost bin, where ample kitchen scraps serve to keep their stomachs full. And I learn that these wood snails have formulaic names—humanly given and unpronounceable, but names nonetheless A letter—Y, R, B for yellow, red, brown, respectively—designates the color of the snail, and numbers, the bands on each whorl. So my plain champagne-gold snails are designated as Y00000, and one of the banded snails is Y12345. If one of the bands is fused to another on whorls 1 and 2, then the formula would read Y(12)345; if a band shows a break between whorls 2 and 3, it would be Y12z345. These names are not, however, unique. Just as a schoolroom may have three boys named Bill and six girls called Ashley, so my garden provides living space for an unknown number of Y00000, Y12345, and variants thereof.

My dark, dank compost bin is also colonized by striped slugs, which creep out at night to dine upon my hostas, hyacinth beans, and seedling vegetables. Like the banded wood snails, they are native to Europe but do not, to my knowledge, have formal names other than the one assigned by Linnaeus in 1758—*Limax maximus*, "the biggest slug." The common names are legion—giant garden slug, great gray slug, leopard slug, tiger slug, and spotted garden slug. But I can imagine that one of these creatures might be distinguished formulaically from every other by the pattern of the spots on its mantle, just as cats can be told apart by the one-of-a-kind arrangement of their whisker-follicles, and people, by their fingerprints. Slug and snail are closely related. Their kingdom, Animalia; phylum, Mollusca; class, Gastropoda; subclass, Orthogastropoda; superorder, Heterobranchia; and order, Pulmonata, are precisely the same. Translated, the terms mean Animals that are Mollusks with Stomach-Feet, Straight Stomach-Feet to boot, with Diverse Gills and also a Lung. Both are Stylommatophora, "slender-eye-bearing stalks" with the eyes at the tip-top of the upper pair of tentacles (the eyes of aquatic snails are situated not at the top but near the base of the optical tentacles). Neither, however, have gills; the name reflects their ancestry. Then the lineage separates. The giant garden slug belongs to the Suborder Eupulmonata, "Good-lung," and the Limacidae, the Slug family. The banded wood snail is a member of the Helicidae, the Helix family, so named for the spiral coils of its shell.

Snails in shells came into the world long before the unclad slugs. Fossil gastropods more than 550 million years old have been found in Cambrian deposits. They were creatures of the sea, breathing dissolved oxygen through gills and counting on the movement of water to unite eggs and sperm. It took a long time for any of them to choose freshwater over salt, much less to venture coming ashore. The early marine ancestors underwent a slow metamorphosis into terrestrial models through several major structural changes. For one, snails living in shallow, anoxic, brackish or muddy environments, like salt marshes, lost their gills and developed a lung so that they could breathe air. Mollusks could not come ashore, though, until they evolved means of internal fertilization and produced eggs that could hold as a food supply for their developing young. Then, while aquatic snails came—and still come—in male or female versions, the land snails doubled up and turned themselves into hermaphrodites. Equally important, they developed the ability to produce their definitive mucus, which protects them from desiccation. By the Eocene Epoch, which began some 58 million years ago, land snails had become abundant. Many present-day genera of aquatic and land snails date back to the Oligocene and Miocene Epochs, 36 to 5 million years ago. Slugs represent a further disportment on the land model, which evolution seems to be pushing in the direction of nakedness.

Anatomically, slug and snail are much alike. The main difference lies in the fact that the snail has a shell, and the slug, a fleshy mantle located atop its body just aft of its tentacles. Snails have mantles, too, but in them it is the organ that secretes the shell. In the giant garden slug, it hides a plate-like proto-shell. The proto-shell in another family of slugs consists of chalky granules. Both banded wood snails and giant garden slugs are decorated, the snail with its bands (or lack of them), the pale gray slug with small dark gray spots on its mantle and shadowy gray stripes down the rest of its body's dorsal side.

The ridge down the center of the slug's back is called a keel, and the fringe around the lower edge of the foot, a skirt. Both snail and slug are truly "stomach-feet," and, except for its head, the body of each is a muscular foot that holds stomach, heart, liver, kidney, a single lung, intestines, genitalia, and other organs. These mollusks move by contracting and relaxing their foot. The tentacles of both snails and slugs come in two pairs, the longer, upper pair equipped with minuscule light-sensitive eyes and the lower with the ability to smell, feel, and perhaps taste. If a tentacle is somehow lost, it will regenerate in a matter of weeks. The mouth, set on the underside of the head, has a jaw and a rasp-like tongue called a radula; and the radula has teeth, which wear down quickly as they press food against the mouth's hard roof. But because this toothed tongue grows continuously from the back of the mouth, new, sharp teeth are always available to replace the old. If you look closely at a slug's mantle, you'll see a small hole on its right side—the pneumostome, "air-mouth," leading to its lung. Both slug's mantle and snail's shell conceal more than they show. The slug's anus lies beneath the mantle, on the right side a tad tailward of the pneumostome, while the anus of a snail is tucked just inside its shell. Headward on each animal, in the place that an ear might be if it had an ear, is the genital opening. The snail and the slug are hermaphrodites that produce both sperm and eggs.

The business of mollusk mucus needs a section of its own, as does the business of mollusk sex. The gummy stuff characteristic of slugs and snails is manufactured in the animal's foot. It plays important roles in the well-being of terrestrial mollusks, for it offers protection from dehydration, aids movement by decreasing friction between the animal and its path, and reduces the risk of injury. It also allows the creatures to climb vertical surfaces and to stick without fear of falling to the lid of my compost bin. Some land snails let several layers of mucus dry to form a door when they are tucked into their shells; such a tactic keeps them from dehydration when they go into winter hibernation or wish to make it through a period of drought. (Some, though not the banded wood snail, are blessed with a door, called an operculum, which they can shut tight when they're curled into their shells.) The entire body of a giant garden slug produces mucus, which acts as a sort of protective shell. Mucus also figures in that animal's lovemaking, of which, more in a trice. And the Limacidae, the family to which my garden slug belongs, makes not just one but two kinds of mucus. One, which might fairly be called "slime," is a thin and watery substance produced by the outer skin of the foot, and the other, thick and more like carpenter's glue, comes from a gland at the front of the foot. It is dried mucus, of course, that makes the shiny, zigzag trails. The trails of slugs and snails may be distinguished one from the other because the former leave an unbroken sign of their passage, while the trail of the latter is interrupted.

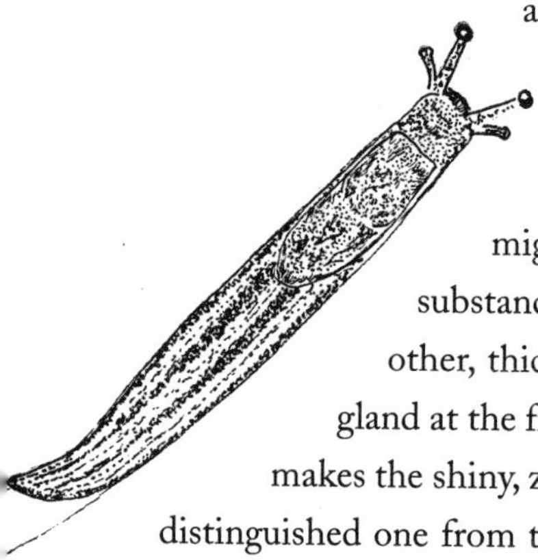

In the matter of snail sex, I have seen the banded wood snails courting, ear to ear as it were, on my front walk. It looks to be a cozy arrangement, as tender as a caress. But it's just the beginning of a complicated *pas de deux* that can last for hours. The point is to exchange the sperm of each so that the eggs of each will be fertilized. The internal organs that participate in this act of conjugation begin with the ovotestis, the gonad that produces both sperm and eggs, which is located in the pinnacle of the snail's whorled shell. It sends its products through a canal called the sperm oviduct, which delivers sperm to the penis and eggs to an area just above the vagina. When conjugation actually occurs, the snails meld themselves sole of foot to sole of foot—you might call them sole-mates—and join their genital pores, through which the transfer of sperm takes place. It is exchanged in the form of a little packet called a spermatophore, out of which the sperm swim into a fertilization pouch. The whole mating process can take hours. In about two weeks, in leaf mold, under a stone, or in my compost bin, each one of the pair will dig a nest with the hind part of its foot and lay its little, beadlike white eggs. Some terrestrial snails—the giant African land snail, for one—are viviparous, retaining the eggs within their bodies until they hatch and tiny, exquisitely fragile snails, complete with shells, emerge into the world. But my garden snails (note how I claim them as mine) are entirely conventional in their egg laying.

Sex between giant garden slugs is not an act that I have yet seen, but photographs of their entwinement show an aerial ballet of stunning drama. Scent likely alerts one slug to the presence of an eager other. The initial moves are vividly described by Mr. Lionel E. Adams, B.A., in a paper, "Observations on the Pairing of *Limax maximus,*" that he read to the Royal Society on December 8, 1897:

> When the pursuer overtakes the pursued, each touches with its tentacles the tentacles of the other, after the manner of ants. Then begins a circular procession, each with its mouth at the other's tail, and this procession lasts from half an hour to two hours and a half.

He then notes that the two engage in a mutual and mutually enthusiastic licking of each other with their radulae. The action stimulates the mucus glands to secrete the sticky type needed for the next step. Let Mr. Adams continue:

> The circle now grows more contracted, the slugs overlapping and showing evident excitement, the mantles flapping before and behind. Then, suddenly, the slugs intertwine fiercely, and launch themselves into space, heads downward, but suspended by a thick strand of mucus for the distance of 15–18 inches.

Like snails, they are sole-mates, and it is as if they have suspended themselves headfirst by a bungee cord firmly attached at the top by a wad of gluey mucus. Mr. Adams says that he has seen couples suspended from many places, including an outhouse beam, the leaves of a currant bush, and the glass pane of a greenhouse, but that the usual place is from a wall or the trunk of a tree. Immediately after the pair is suspended, each member slowly extrudes its penis, which first emerges looking like a club. Each club, like a seed unfolding a single cotyledon, expands and puts on a frill around its edge, and the two join in an arrangement that resembles an upside-down two-stemmed cocktail glass of a glistening blue slightly paler than the Bombay Sapphire gin bottle. The pair, in a process that may take two hours, is exchanging sperm. After the exchange has taken place, each mollusk retracts its penis and tucks the sperm received into its spermatheca, its sperm-vault, whence it moves into a special pocket where the eggs are fertilized. One member of the pair then drops to the ground, and the other ascends the mucus bungee cord, eating it on the way up. Eating it makes great good sense: waste not, want not. To produce it demanded a considerable expense of energy, and to consume it (much as a spider consumes the silk of an old and tattered web) constitutes a means of recycling the protein that it contains. Eggs will soon be laid in a sheltered place beneath leaves or, of course, in the compost-bin debris; eggs released in the fall will overwinter to hatch in balmier times.

Nonetheless, no matter how glorious and nutritious its sex life, *L. maximus* has bad habits other than eating my hostas and tender vegetable seedlings. It has been included in a list of "cellar mollusks" compiled by the same Dr. W. G. Binney who imported more than a hundred banded wood snails into Burlington, New Jersey, in 1857. Not incidentally, a genus of slugs is named for him—*Binneyi*. Here's what he has to say about the nighttime forays of the giant garden slug and the two other kinds of slugs that find the basement an ideal, temperature-controlled home that lets them prowl year-round instead of hibernating during cold spells and aestivating in summer drought. Dr. Binney is, mind you, talking about the cellars of the nineteenth century, which stored not only root vegetables but other foodstuffs, as well as kitchen scraps.

> Our cellar mollusks are all nocturnal in their habits. They lie quietly stowed away in some crack or crevice of the walls during the day. At night they sally forth in pursuit of food and to enjoy the company of their kind. They feed on vegetable matter—refuse from the kitchen, decaying vegetables or fruits—or on Indian meal, flour, or anything they are lucky enough to find. They even devour animal food, and in confinement have even been accused of cannibalism.

The giant garden slug is a cannibal not just in confinement but also in the wilds, though not necessarily of its own kind. *L. maximus* is a mugger, as well, zeroing in on native slug species and leaving them to die unshriven at the scene of the attack.

The only attacks that concern me are those wrought by both snails and slugs on my gardens' floral and vegetable greenery. (It's not important that they also eat the apple slices used for bait in my Havahart trap.) So, I can counterattack, choosing from a menu of tricks to control the damage. One is to rake the gardens removing leaves and other debris under which the sneaky creatures hide. The rake may also gather in over-wintered snail and slug eggs. Another trick is to water the gardens in the morning so that the plants are dry by evening when the mollusks come out to dine. Yet a third trick is to put in plants that do not appeal to the mollusk palate. Shun primroses, lobelia, hyacinth bean, strawberries, and lettuce. Embrace daylilies, yarrow, periwinkle, and phlox. The problem is that, while it's easy to embrace many plants, especially daylilies, it's almost impossible to shun vegetation that mollusks find delectable. What would spring be without lettuce and strawberries? Summer without a hyacinth bean climbing a trellis in the front garden and charming both me and passersby with its clusters of pale purple flowers?

Reducing mollusk depredations might also involve trapping slugs by laying out a layer of wet newspapers or a cantaloupe rind, checking beneath it the next morning, picking up the hidden slugs and snails by hand, and plopping them into a jar of sudsy water, where they drown. By hand? Yes, shielded by disposable plastic gloves. Then, beer attracts mollusks. I could make a beer bar by sinking a jar, can, or cottage-cheese carton into the ground and filling it to the halfway mark. A water-and-yeast mixture would also suffice, for it's the odor of fermentation that beckons to the garden mollusks like a siren song. Another stratagem could protect my raised beds: installing copper tape or strips, which emit a small but discouraging electrical charge. Lethal commercial slug baits exist, but I eschew them, for disposing of slugs killed by chemicals is not a simple matter—their bodies might contaminate a water supply or poison an animal that eats the carcasses. I choose instead to abrade the mollusks' soles, a tactic that causes dehydration and death. Several irritants could be used—wood ashes or crushed eggshells—but my abrasive of choice is diatomaceous earth, silicon dioxide, made of the fossilized shells of ancient marine microorganisms. It is fine, gritty white stuff. Plants that were full of holes are setting out new, intact leaves. Reinvigorated, the hyacinth bean starts climbing its trellis, and the season's first strawberries turn red as I write.

But I cannot win. I cannot do away with all the mollusks in my garden, not by a long shot. They, too, employ the safety-in-numbers strategy of mouse and turtle and reproduce with fine exuberance so that their species will continue even if a multitude of individuals is lost. What to do? Plant enough for them and me.

THIRTEEN

Mr. McGregor's Nemesis

{ *A Story About Wool-Tails* }

"I can't grow anything!" Ethel said last year. "The rabbits eat it all." It was her beloved flowers that had fallen victim to long incisors. And the cottontails in her yard defied counting. They came in all sizes from hold-in-the palm-of-your-hand to fully mature, and they could be seen nibbling away at dawn and at dusk, with occasional depredations in between. Despairing, she asked me what to do. The overgrown yard of a neighbor—volunteer sumac trees, bush honeysuckle, and other shrubs, weedy undergrowth, and poison ivy—borders hers; the rabbits have ample cover. The only advice that I could give was to build a fence and bury it at least a foot in the ground. She sighed, as frustrated, as downright defeated, as the legendary Mr. McGregor.

My own encounters with cottontails have been less aggravating. Looking back several years, I remember a spring and summer in which the rabbits were legion on my street. Daily, I saw cottontails—large, medium, and small—in the backyards adjoining mine. I saw them both awake and snoozing on my neighbors' front porches and saw them hopping lippertylippertyup and down the blacktop. Only rarely did I spot one in my yard, even though the gardens offered a bounty of greens. (My scourge was, is, and always shall be the groundhog.) But during the following year and the year after that, rabbits were nowhere to be seen in my neighborhood. An absence, however, does not mean that they won't return when they are good and ready.

Looking back several decades, I recall the very young, very small rabbit that the cat brought in and laid in triumph at the feet of my elder daughter, Lisa. Of course, she picked it up and inspected it. The little animal had suffered no injury. But it was not returned to the great out-of-doors. Instead, naming him Jeremiah (though we never knew the rabbit's gender), she installed it in one of the many cages in the basement. The reason that cages—hamster cages, guinea pig cages, birdcages—occupied much space in the lower regions of the house is that Lisa had recently completed a ninth-grade science project in which she used white rats to test the efficacy of several over-the-counter products, like No-Doze, designed to keep people

awake. A dozen rats soon became six dozen rats, but that is another story. Jeremiah did not die of shock as do many wild rabbits when they are taken captive. Rather, the small creature thrived and in three weeks became a middle-sized cottontail.

I was dubious, however, about the wisdom, not to mention the legality, of keeping a wild animal caged as a house pet. It took three weeks, but Lisa finally yielded to my pleas that the little rabbit be released. The cat brought it back the very next day. This time it had sustained fatal wounds. I would like to think, but don't know for sure, that we encoffined Jeremiah in a shoebox and buried body and box with tender ceremony. Cottontails live for an average of a year in the wild, although, with luck, they can survive as long as three years. To this day, I wonder what we should have done when the cat presented his catch to his person. To have let the cottontail go right then would have invited the cat to continue his play. To keep it violated both the law and common sense. No wildlife-rehabilitation center was at hand. Whatever we did, Jeremiah was doomed.

The second cottontail that comes to mind was never given a name, but she was a familiar presence for a whole summer in the days that my husband, the Chief, and I gardened and fished in North Carolina. Again, I never learned the gender of this particular rabbit, but it's fair to assume that the animal was a doe because of her size—does are larger than bucks. As cottontails go, she was immense, fully as big, though not so heavy, as a toy poodle. She would graze afternoons on the wild lettuce that sent up deliciously leafy stalks in the backyard; then she'd lie down and take a nap. Our presence, our movements did not disturb her. And she was old. Not only was she large, but she was half-blind; one of her eyes was glazed over with a milky cataract.

The other rabbits with which I've been peripherally associated were not cottontails. They were satin rabbits of a rich red color that my younger daughter, Hannah, bred and displayed at rabbit shows. They garnered hundreds of ribbons—white, yellow, red, blue, purple best-in-show—and trophies enough to fill several bookcases. Her satins and every other domestic breed in the world are descended from the European rabbit, *Oryctolagus cuniculus*, a binomial that combines a Greek genus name meaning "digging-hare" with a Latin species name that translates simply as "rabbit." The original range of the European rabbit was southwestern Europe and North Africa, but it followed the vegetable patches as agriculture moved north and west. Aside from reproduction, digging is what it does best. The primal, still wild European rabbits are the warren-digging, loquacious, beleaguered animals of Richard Adams's *Watership Down*. Such warrens provided early golfers in Scotland with hazards, in which a ball was often lost forever. These animals are also the rabbits that overran Australia after their introduction in 1859. In the domesticated versions of the European rabbit, some are pets and show animals; some, like angoras, are raised for their soft wool, others with their velvety skins, for the fur industry; and still others, especially the New Zealand and Californian breeds, for meat. And what a motley crew the domesticated animals are! They represent genetic tinkering at its most inventive—dwarfs and giants, rabbits with tumbledown lop-ears, checkered rabbits, spotted rabbits, longhaired breeds, rabbits with a white blaze down their faces, and other extravagant variations on the general theme. And the array of colors amazes: white, red, black, gray, blue, tortoise, chocolate, lilac, gold, fawn, and more. The dry fecal pellets of domesticated rabbits kept in pens make for a garden fertilizer as potent as chicken poop or horse manure. But those given free run of a house can be trained to use a litter box.

But no one has succeeded in domesticating Eastern cottontails, though, like Jeremiah, some have been kept captive. They will stay wild, no matter how much loving attention is focused on them—aside from which, a cottontail can deliver a punishing kick when it has a mind to. The old doe that napped in my North Carolina yard, the ravenous animals that ate Ethel's flowers into oblivion, and the other members of the species are formally known as *Sylvilagus floridanus*, "forest-hare from Florida." Florida? Yes, because that's the place in which the type specimen was first identified and described. And *lagus*, the classical Greek term for "hare," is a catchall word that encompasses both hares and rabbits, for the latter were—and still are—very much present in Greece. In the U.S., it's common to use the words "hare" and "rabbit" interchangeably, but the two kinds of animals are easily distinguished one from the other. Newborn rabbits, like new-hatched wrens, are altricial, coming naked and blind into the light, while hares, like ducklings, arrive with their eyes wide open and their bodies fully clad in downy hair.

The genus *Sylvilagus* encompasses seventeen species. They all belong to the kingdom Animalia; the phylum Chordata; the class Mammalia; the order Lagomorpha; and the family Leporidae. That all unravels as Animals with Backbones and Mammary Glands that come in Hare-shapes and belong to the Hare-and-rabbit family. A third kind of Hare-form exists in addition to the hares and rabbits: the pika, an attractive little animal of the genus *Ochtona*, which has some thirty subspecies living mainly in far northern regions; the genus name is a Latinization of the Mongolian word for the animal *ochodona*. No matter what the species, all look like miniature guinea pigs with light gray fur and small round ears. The North America species live in high, cold, rocky places, which are now shrinking because of climate change, and with the disappearance of habitat, pika populations also shrink. They have been put on the vulnerable list by The World Conservation Union (IUCN). We are not in danger, however, of losing most of our leporids—our hares and rabbits, that is. As with the European rabbit, the IUCN rates cottontails as animals of least concern, occupying the capacious bottom of the pyramid that peaks with endangerment and ends with extinction.

And, oh m'goodness, the common Eastern cottontail, *S. floridanus*, boasts at least thirty-four subspecies, the least number of which of which live in North America. The rest are found south of the border from Mexico into northern Colombia and Venezuela. Some names assigned to subspecies specify a habitat preference: *S. f. aquaticus*, the "water" cottontail and *S. f. paulustris*, the "marsh" cottontail. Other names often indicate a geographic point of origin or populousness, as with some Central American rabbits: *S. f. aztecus*, the "Aztec" cottontail; *S. f. hondurensis*, the "Honduras" cottontail; and *S. f. yucatanicus*, the "Yucatan" cottontail. One of the South American subspecies is *S. f. supercialiaris*, the cottontail "with eyebrows." Jeremiah, the old doe, and Ethel's adversaries all belong to *S. f. mallurus*, the cottontail with a soft "wool-tail."

Though no one yet knows for certain, the earliest Hare-shapes—the lagomorphs, that is—seem to have come scampering onto the world-scene in northern Asia some 58 million years ago as the Paleocene Epoch was sliding into the Eocene. That was the time in which plants, the gymnosperms and angiosperms, began to flourish in earnest, thereby providing rabbitkind with all sorts of succulent green foodstuffs. Fossil evidence shows that the early lagomorphs came from much the same mold. Evolution produced only few real variations on the general pattern. By the time that the Oligocene rolled in around 37 million years ago, true rabbits were doing the bunny hop and gobbling up vegetation in both the Old World and the New. Some of the transformation can be charted in skulls, jaws, and dentition, and dentition is prime, for teeth survive the eons far better than any other body part. Lagomorph teeth, not just the incisors but the molars as well, are open-rooted and grow throughout an animal's life. So do the teeth of rodents, a fact that long caused scientists to conclude that the leporids—the rabbits and hares—were indeed rodents, like squirrels, groundhogs, beavers, porcupines, and mice. Although in the earliest days, rodents and lagomorphs probably shared a common ancestor, the much more recent fossil record shows distinct lines of development, especially in regard to the arrangement—not the constant growth but the arrangement—of their teeth. Lagomorphs possess four upper incisors, while rodents have only two.

With rabbits and hares, the second set of incisors, which do not keep growing, snuggles behind the first set. A single layer of enamel covers lagomorph teeth, but two layers bedeck rodent teeth. To keep these ever-emerging teeth in trim, the both rodents and lagomorphs are given to gnawing (and pet rabbits kept in the house will gnaw everything from woodwork to electric cords). But as a rule, all lagomorphs are vegetarians, while rodents eat meat as well as greens. And rabbits and hares don't walk; evolution has elongated their hind limbs and strengthened their muscles, which requires them to hop and bound in order to get from hither to yon. The little pikas, however, still scamper. The features that distinguish lagomorphs from rodents do not stop here. Unlike mice, squirrels, and all their kin, rabbits and hares are built with what might seem an exceeding odd configuration: the testicles are placed in the groin on either side of the penis. Then, there's the matter of the penis bone. Most mammals, including rodents, have an *os penis*; so do marsupials, though in the opossum, at least, it's made of cartilage, not bone. The creatures that lack such an anatomical feature are human beings, the anthropoid apes, and the lagomorphs.

Lack of a penis bone hardly stops cottontails and all the other leporids from reproducing explosively. Female cottontails may become sexually mature at three months of age and give birth to as many as four litters a year. Although no one knows for certain when it comes to cottontail behavior, it seems likely that they, like the European rabbit, come into heat but must copulate in order to trigger ovulation. Gestation takes about four weeks. A female will build a nest, called a form, in a depression on the ground and line it with grass and some fine, soft fur from her belly. She'll kindle her kits and mate immediately thereafter. She is a come-and-go mother, nursing the babies only once a day—she must, after all, eat almost nonstop during her waking hours—but her milk is obviously nutritious enough to keep the young ones thrifty. By the time that they are sixteen days old, the kits are wide-eyed, furry, and able to fend for themselves. Why is the world not filled to overflowing with rabbits? Because their lives are short, and most fall victim to domestic cats, dogs, coyotes, raccoons, hawks, crows, and other carnivores, including us, before their first year is up. Their populations also rise and fall with climate conditions and the abundance or scarcity of food. The overrunning of Australia by the European rabbit came about because native predators were lacking.

Cottontails and other lagomorphs do not hibernate. Summers, they gorge on fresh greens, particularly those that are newly sprouted, but in the winter, they depend on bark, twigs, buds, berry canes, and even poison ivy vines. All herbivorous animals, eating plants loaded with cellulose, require special sorts of digestive mechanisms. Two strategies have been developed, both employing fermentation. The strategy used by cows, sheep, goats, and other grazing or browsing ruminants is the four-chambered-stomach strategy. Grass goes into the rumen, a fermentation vat, where bacteria break down the cellulose. The next step is the reticulum, which lets the animal regurgitate its food as a cud and chew it to break it down further. The food then travels to the omasum for more processing and thence to the true stomach, the abomasum, which breaks down the nutrient-bearing bacteria with a digestive enzyme. The second strategy, that of horsekind, cottontails, and the other lagomorphs is called cecal digestion. Cecal digesters put part of their large intestine, the cecum, to work as the fermentation vat. With horses, the process proceeds pretty much as you'd expect: food goes in one end and comes out the other.

With cottontails and other rabbits, the process is reduplicated—it happens at least twice, that is—before fermentation can release from the food the nutrients that the animal needs. Their cecum is huge, far larger than the stomach. And in this vast chamber, bacteria and other microbes break down plant matter to extract vitamins and proteins. Soft green feces result from this first go-through, and the rabbit eats this version of a cud. On round two, the stomach digests the recycled food and sends it on to the small intestine, where nutrients are absorbed. But the leavings from the second round do not necessarily constitute the hard, round pellets that rabbits excrete. Those firm brown pellets consist of the leporid equivalent of cherry pits, the tough outer skin of corn kernels, and other things that people can't digest. The soft feces may be used over and again until all their goodness has been used up. And the rabbit, nibbling away at its vegetables, keeps giving its cecum more cellulose to process. If the disgusted owner of a pet rabbit prevents it from ingesting its poop, the rabbit may very well become malnourished and die.

*

What are cottontail rabbits good for? They certainly merit a place in the pest category as far as Ethel and other gardeners are concerned. Their populations, though still sizeable, are not so large nowadays as they were one hundred years ago. One reason is loss of habitat. The animals like swamps, woods, thickets, and brush piles, places that offer protective cover during the day. But the brush that grows up in logged areas becomes shaded out by new trees; industrial agriculture uses herbicides and heavy-duty equipment to keep fields clean; and urban sprawl encroaches on prime rabbit territory. Nonetheless, cottontails reproduce successfully enough to make sure that the Americas do not run out of them. They also provide a steady food supply for other animals, including us. And we like their soft fur, their puffball tails, and the sport that they provide. State wildlife and game commissions offer tips on creating rabbit-friendly habitats—planting food plots of clover and grasses, creating cover with artificial brush piles, and clearing forest edges so that protective brushy shrubs grow up. Rabbits, not deer or squirrels, are the most popular game animal in the U.S. And what do you do with a wild rabbit that you have bagged? Cook it and eat it, of course.

Without the New World's cottontails and other leporids, the early colonists would have been even hungrier than they were as they made their way into a perilous *terra incognita. American Cookery*, the first American cookbook, written by Amelia Simmons and published in 1796 in Hartford, Connecticut, has much to say about rabbits. It pronounces wild rabbits as the best to eat and advises using your nose to determine the freshness of the kill. Thinking beyond the mere consumption of rabbit meat, she writes, "The cultivation of Rabbits would be profitable in America, if the best methods were pursued—they are a very prolific and profitable animal." She knows, however, that the wild cottontail is not a candidate for profitable cultivation, for she speaks of a would-be cultivator spending money on "a Rabbit's borough." The cottontail, no matter how delicious, does not dig a burrow. But, oh, it has inspired a mouth-watering multitude of recipes. The authors of *The L. L. Bean Game & Fish Cookbook* say, "Rabbit can be prepared with *any* recipe that you use for chicken. In my humble opinion there is only one difference: rabbit is better than chicken. And I love chicken." They then proceed to give a blessed plenty of gourmet recipes, including rabbit stew, Maine style; rabbit in coconut milk; curry of rabbit; and rabbit baked in tarragon, mustard, garlic, and cream.

*

We live in a world inhabited not just by real rabbits but also a nest of virtual rabbits.

"Nest" is an old collective term for a group of rabbits, even though cottontails tend to be solitary—except when they *are* newborns in a tangible nest. ("Husk" is the collective word for hares.) Among our imaginary rabbits, we can count the Easter Bunny, the White Rabbit who leads Alice down the rabbit hole to Wonderland, Bambi's friend Thumper, and a great gang of cottontails—Mr. McGregor's nemesis Peter Rabbit and Peter's siblings, Flopsy, Mopsy, and Cotton-tail; Peter's cousin Benjamin Bunny; the trickster Br'er Rabbit in his briar patch; the wise-cracking Bugs Bunny, another trickster; and the Playboy bunny with her skimpy costume and outsized puff of a tail.

Rabbits are involved in magic, too. Carrying a rabbit's foot has long been thought to bring good luck. And I still follow a ritual practiced by my family, a ritual that comes from some obscure nook in folklore's vast cabinet: the first word spoken on the first day of each month must be, "Rabbit!" To do so ensures good fortune for the next few weeks. Some of my cousins vie to be the first to make a long-distance "Rabbit!" call. But if, somehow, something else should slip out, like a wee-hours plea that your spouse stop snoring or a reprimand to the pet leaping onto the bed, bad fortune would follow sure as the moon waxes and wanes until the new month brought a fresh opportunity to invoke benignity from Lady Luck. In my teens, I elaborated on this formula by turning it into three triplets, three being a powerful number, as with the Trinity or the Three Bears: "Bunny, bunny, bunny, rabbit, rabbit, rabbit, hare, hare, hare." That way, all bases would be covered. But age has brought moderation.

*

I doubt that Ethel would find any use in practicing the "Rabbit!" ritual. But a year has passed, and the gardening season flourishes. I've been digging, sowing, transplanting, and waging my annual war against the groundhog. But I hesitate to mention the word "garden" to Ethel. So, I offer commiseration. "Are the rabbits plaguing you again this year?"

"Haven't seen a one," she says. "My garden is just *great.*"

She profits from a winter that was hard on rabbits. So, most likely, did Mr. McGregor. But they'll be back. Dem bunnies gonna rise again.

FOURTEEN

The Harlequins

{*A Vampire Story*}

Late summer, and the horseradish plants' bright green, ruffled, banner-like leaves, some of them nearly a yard long, are liberally sprinkled with pale yellowish spots and splotches. The vampire has been at work, sucking out the plant's sap and chlorophyll, which are its lifeblood. I know the vampire's name—harlequin bug—and not just one but a slew of them have stationed themselves on the leaves like boldly bright

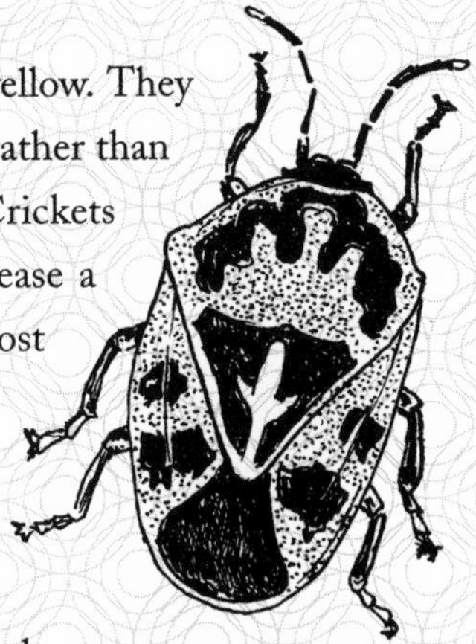

confetti—black and red-orange with a hint of yellow. They are a bit larger than ladybugs, but shield-shaped rather than round. How silently they go about their work! Crickets chirp, mosquitoes whine, gorging caterpillars release a rain of droppings. The thirsty little vampires are most curious creatures, well worth contemplating.

Sometimes they're called calico bugs, calicobacks, or firebugs. But the most common of their common names is harlequin bug, and that name fits like a perfectly tailored, quite theatrical costume. And, indeed, the pattern on the bug's back, which looks like an X with four truncated diamonds meeting just below the center of the shield, replicates the gaudy red, yellow, and black diamonds in the suit of Arlecchino, a buffoonish but lustful servant-character who's been around since at least the mid-1500s in the Italian Commedia dell'Arte. And the bug's little black head resembles Arlecchino's black mask. The bug's formal name, *Murgantia histrionica*, states its connection with the theater, for *histrionica* means "actor." *Murgantia* remains a mystery.

The bug was first described and named in 1834 by the German entomologist Carl Wilhelm Hahn (1786–1836), who is, alas, no longer able to explain the reason for assigning a most peculiar genus to the creature.

The classification of the harlequin bug goes this way: Animalia, Arthropoda, Insecta, Hemiptera, Pentatomidae, or—put into plain English—Animal kingdom, Joint-Foot phylum, Insect class, Half-Wing order, and Five-Section family. Joint-Foot needs no explanation. Half-Wing characterizes the double nature of the forewings (which are certainly whole), for at their base, where they join the body, they are leathery, while the ends are membranous. The hind wings, shorter than the forewings, are completely membranous. The order Hemiptera is that of the true bugs, which run the gamut from cicadas and water striders to aphids and bedbugs. The most significant feature of all Hemipterans is that their mandibles and jaws have joined to form a proboscis protected by a modified lip, and this beak is used like Dracula's canine teeth to pierce plant tissues so that a plant's vital juices may be sucked up. As for the Five-Section family, the name refers to the five segments of its members' antennae. And those that belong to the sub-family Pentatominae are, for good reason, collectively known as stink bugs. They come equipped with a chemical factory—a pair of scent glands—in their thorax that emits acetate, butenal, and other noxious substances. They are not only sap-sucking vampires but also the skunks of the insect world, using chemical warfare to thwart would-be predators.

Harlequin bugs, though dressed like buffoons, are hardly as slow-witted as their comedic namesake, and judging by the numbers that bedeck my horseradish plants, I reckon that they are far more successful than Arlecchino in their amours. The way in which they manage to replicate themselves is truly curious. The eggs of all stink bugs are shaped like miniature kegs with straight sides, slightly domed lids, and decorative markings. They are laid in tidy rows of twelve to fifty, and a single female will deposit one batch here, another there, and yet another nearby, until all her eggs have been emptied out. The French entomologist Jean-Henri Fabre gets carried away when he describes the cask-like eggs of the cabbage-loving stink bug that he calls the Ornate Pentatoma (now called *Eurydema ornatum*, the "ornate broad-body"). In the elaborate markings on its shield, it much resembles its harlequin-bug counterpart in North America.

> The microscope shows us a surface engraved with pits, like those of a thimble, arranged with exquisite regularity. At the top and bottom of the cylinder there is a broad dull-black band; on the sides is a wide white belt with four large black spots symmetrically placed. The lid, surrounded with snow-white filaments and edged with white, swells into a black dome with a central white spot. In short, a funeral urn, with its

> violent contrast of coal-black and creamy white. The Etruscans would have considered it a magnificent model for their burial vessels.

Except that the little casks contain life's beginnings, not its end. Harlequin-bug eggs have their own distinctive embellishments. The kegs look as if their pale gray or light yellow staves are held together with a narrow black hoop near the bottom and a slightly wider black hoop near the top. As is only fitting for a keg, a bung is placed just above the lower hoop. A small black dot decorates the center of the lid. These little kegs, each about a millimeter high, are deposited, usually in two neat rows, on the leaves of the harlequin's favorite plants. Though some stink bugs suck up the juices of any plant in reach, harlequins are choosy, selecting only members of the Brassicaceae, the Cabbage family, for sustenance and egg laying. But the Cabbage family is an extended one—broccoli, Brussels sprouts, cabbage, cauliflower, collards, horseradish, kale, kohlrabi, mustard, radishes, and turnips, to name only the best known. If cole crops are not available, the bug has been known to suck the living daylights out of such comestibles as tomatoes, eggplants, beans, asparagus, and even fruit trees. So, in any garden, the harlequin's larder is full to bursting, and space abounds for its nurseries.

New harlequins emerge from their leaf-cradled kegs after twenty days in early spring but after only four or five when summer and hot weather arrive. According to Fabre, who observed the process closely, hatching is an astonishing event. And it is the same for all stink bugs, be they ornate, brown, green, harlequin, or other species. After the hatchlings had left their casings, he noticed a peculiar shape within each egg near its rim: "a black mark in the shape of a broad arrow." At first, he thought that it might be a "trademark," a bolt, or a hinge to hold the lid shut until the hatchling opened it. It's none of these. Rather, it's a device, a kind of three-pronged, pointed cap, with which the emerging bug has been equipped just for the task of prying open the lid of the keg—an arduous task, a truly Herculean task, when we take into account the fact that the new bug is a soft, squish-able creature, quite bereft of any armor. It is the equivalent of a bird's egg tooth or, for that matter, a turtle's. The moment for hatching arrives; the cap is at the top of the bug's head. Fabre tells what happens next:

> Look carefully, and there, involving a certain small area, almost a point, you will see rapid pulsations, we might almost say piston-strokes, produced, beyond a doubt, by sudden waves of blood. By hurriedly injecting what little fluid its body contains under its pliant cranium, the tiny creature turns its weakness into energy. The three-cornered helmet rises, pushing upwards, always pressing its points on the same point of the lid.

Exerting relentless pressure, the buglet lifts the lid at an angle and at last, after more than an hour, creates an opening big enough for exit. At that point, it sheds its cap. Then the new-hatched bug unfolds its antennae and its legs and moves into the world, although not very far at first. It stays near its keg until its body gains some firmness. Before it reaches maturity, the buglet, now called a nymph, must go through five instars—five sheddings of its skin, that is. As a nymph, its coloration suggests the gaudiness to come, for its basically black body is adorned with symmetrical reddish spots. After six to eight weeks of swilling down the juices of Cabbage-family plants, it will not only don Arlecchino's yellow-accented red-orange and black outfit but also, in the form of wings, get its license to fly.

Rumors have attended stink bugs. Fabre cites a passage from the work of Baron Karl de Geer (1720–1778), a Swedish entomologist, who maintained that the mother stink bug herds her young from food source to food source and leads them under a leaf when it rains. Not only that, but she protects them vigilantly from predators, especially their father, who is inclined to consider them dinner. Not so, for how would the mother be able to locate all of her offspring, the eggs having been laid in various foliate places? Aside from which, most, though not all, stinkbugs are vegetarians. I have, however, seen the nymph of a brown stink bug (*Euschistus servus*, the "well-split, slavish" bug) practicing carnivory right out in plain sight while I was picking lima beans in my North Carolina garden: its proboscis was up to the hilt in a caterpillar.

The second rumor has to do with the way in which the harlequin arrived in the United States, where it can now be found from the Atlantic seaboard to the Pacific and as far north as New England and the Great Lakes. They are far more pestiferous in the hot, humid south and can cause damage on a huge scale not just to vegetables in home gardens but also to commercially grown crops. But they didn't originate in North America. Rather, they are native to Central America, and Mexico furnished the specimen on which Herr Doktor Hahn based the type description. How did they make their way north? By human means, as so many other immigrants, many of them stowaways, have entered the U.S. But according to the tale, harlequins did not hitch a ride in the cuffs or pockets of an unwary traveler, or in the Cabbage-family plants that he was carrying. No, their introduction is supposed to have been intentional: Union soldiers turned them loose in Rebel country to devastate the cole crops and bring a concomitant belly-growling despair to Southern troops. The Union was safe from depredation because the harlequins prefer to dine below the 40th parallel.

Aside from the horseradish, I grow other Cabbage-family crops, mainly broccoli, cauliflower, and radishes. But for reasons known only to themselves, the little vampires have never thirsted for the juices of anything other than my horseradish. The victim is worth a few words. It's *Armoracia rusticana*, "rustic horseradish," a native of the Mediterranean region. The "horse" in its common English name is a corruption of the word "coarse," though some say that "horse" came from the German word for the plant, *Meerrettich*, "sea radish," in which *Meer* was heard as "mare." The early uses of horseradish in its home territories seem to have been primarily medicinal. The Roman natural historian Pliny the Elder (23–79) mentions it saying that its heat takes away "scabs, scares, and manginesse." The plant made its way to northern Europe certainly no later than the Renaissance. There it served both as a pharmaceutical and a condiment. The sixteenth century British botanist John Gerard lists its abilities to heal a slew of illnesses:

> The leaves boiled in wine, and a little oile olive added thereto and laid upon the grieved parts in manner of Pultis, doe mollifie and take away the hard swellings of the liver.
>
> It profiteth much in the expulsion of the fecondine or after-birth.
>
> It mittigateth and asswageth the paine of the hip or haunch, commonly called Sciatica.
>
> The root stamped and given to drinke, killeth the wormes in children.

From what Gerard says, the English ate horseradish to cure their ailments, but he acknowledges that the Germans were in the habit of grinding the roots for sauce and using it on meat much as the English used mustard. (To this day, Germans distill a horseradish schnapps.) One hundred years after Gerard's day, the Englishmen, albeit of the lower classes, did eat horseradish as a condiment, and it crossed the Atlantic with early settlers. Gerard grants, however, that horseradish improves digestion far better than mustard. Strangely, though, he says nothing whatsoever about the difficulties of grating those roots.

Grating them is a task of torrentially eye-watering, nose-running proportions, much more so than grating an onion could ever be, for the roots are loaded with allyl isothiocyanate—that's the fancy term for mustard oil. I'll dig and clean the roots, slice them in quarter-inch rounds, and put them in the food processor. I can make about four pints—a meld of roots and white vinegar with a dash of salt—before the pain sets in, before the tears completely blur my vision. But, oh, the pain and weeping are worthwhile: horseradish to be mixed with cottage cheese, to slather with sour cream on roast beef, to give bite to a Bloody Mary. The presence of vampires on the leaves has not—hurray!—affected the roots of my plants, for the little suckers did not start their ravishments till mid-summer. By then the roots had received all the nourishment that they needed.

Live and let live. The relationship between me and the harlequins is neutral.

There's enough horseradish to keep all of us happy.

FIFTEEN

A Gift from a Ghost

{*The Story of Music in the Backyard*}

My backyard is a concert hall from May's beginning well into July, and the main stage is the Norway spruce. Cardinal, catbird, house wren—all perch on green branches that wear long, lush fringes of down-drooping needles. There, courting and claiming territory, the birds sing their hearts out. The catbird renders an almost *sotto voce*

collection of little tunes, each one running somewhat creakily right into the next. The cardinal's *birdy-birdy-birdy* and *cheer-cheer* are answered by his mate, who is an equally accomplished vocalist. The small cinnamon-brown wren begins his song with a breathy introduction that rises and bursts into bright music before it falls and ends. And as he sings his courting songs, he lifts his chin, points his delicately barred tail straight at the ground, and flutters his wings rapidly. Nor is there much pause between one performance and the next. The two other species build their nests in the deep dark recesses of the Norway spruce, but the pair of wrens uses the birdhouse that was put up three years ago, suspended from a bracket that's attached to a stout six-foot-high post.

The birdhouse was given to me by a ghost. A few years ago, when I was translating Virgil's *Georgics*, a long Latin poem on farming in Italy during the first century B.C., I used an old atlas to locate the places—the countries, towns, rivers, and mountains—about which the poet wrote. On the flyleaf of the atlas were a signature and a date written large in a lady's Spencerian hand: *S. Bess Summerson, October 19, 1894.* S. Bess, whose given names were actually Sarah Elizabeth, was my grandmother's sister. How I came by the atlas, I do not remember, but most likely, being interested in classical myths and languages, I spotted it in my grandmother's bookcase and just helped myself. I may have asked permission to keep it, and, then again, I may not. My great-aunt died before I was born; but to see her signature, inscribed when she was sixteen or seventeen years old, waked my curiosity. When I mentioned her to one of my brothers, he not only gave me her married name but said also that he thought some of her grandchildren lived on the other side of the Blue Ridge only forty miles away. What would we do without the Internet white pages? There, I found four families with her surname, along with their addresses and phone numbers. My newfound second cousins and I quickly arranged a grand get-together at my house, which led to another gathering on their side of the mountains. Before my brother and I left to go home, we were each presented with a small, wooden birdhouse, painted pale gray, that had been made by one of the cousins. It was installed on a post near the Norway spruce in the backyard within two days and soon thereafter claimed by a pair of house wrens. I look at it frequently in the spring and imagine a celestial smile on Bess's face.

But in the birdhouse's second year, I made a mistake. It had been such a delight to have resident wrens that I put up another birdhouse some ten yards away. End of singing. The little wrens are fiercely territorial and will not allow themselves to be crowded. They abandoned the birdhouse. A few lone twigs poked out between walls and floor. I know because, after silence made it obvious that the feathered sprites had moved, I took the house down and cleaned it out. Where it had been chockfull of twigs and a lined cup the year before, only a few small sticks had been placed on the house's floor. Lesson learned: the wrens shall have no competition, nor reason to suspect it.

But what a rickety-looking nest the male builds to use as an inducement to his mate! It is he who initiates construction by piling up twiglets inside the house and, likely, in other suitable cavities nearby, for the species likes to nest in snug, closed-in places. He isn't fussy and might opt not just for the house but also for a flowerpot, one of the old mud shoes on my back porch, and the pocket of my gardening apron. He might even evict another pair of birds from their cavity-nest by putting their eggs under the hammer of his bill and smashing them. The multiplicity of building sites is part of his courtship. His mate will either choose one or turn up her bill at all of them, whereupon he must start from scratch. Once, she has made up her mind, she completes the process by lining his randomly piled sticks with soft materials like feathers, hair, and grasses that will cushion her five to eight eggs, which look reddish because the white shells are so thickly speckled with rust-colored spots. I feel fortunate that they have chosen Bess's birdhouse three years in a row—fortunate for more reasons than his bubbly bursts of song, which he sings not just in courtship but also while she works at her task of incubating the eggs.

House wrens are named for that song and for their instinctive yen to nest in holes: *Troglodytes aedon*. Louis Jean Pierre Vieillot (1748–1831) was the ornithologist who so dubbed them. A native of France, he fled to America during the French Revolution and proceeded to play Adam, giving names to the birds that he saw in the West Indies and the United States. Twenty-six of the genera that he established are still in use; one is the genus *Vireo*. The genus name that he conferred upon the house wren means "cave-diver," which is shared by the tiny winter wren, *T. troglodytes*. The species designation is the Greek word for "nightingale." In Greek myth, Aëdon envied her sister Niobe because Niobe was mother to six daughters and six sons, while she had borne only one child, Itylus. But when her skewed passions impelled her to try murdering Niobe's eldest son, she killed her own son by mistake. Hearing her cries of distress, Zeus transformed her into a nightingale. The playwright Aeschylus describes her in *Suppliants*. Anyone who hears the Suppliants' heartfelt lamenting at being sent into exile from their homeland

> will think that he listens
>
> to her who was wife and now
>
> sings out heart's darkness . . .

Barred from her nest in the green leafrivers
        she trills strange sweetness lamenting her exile
and the notes spill old tears with new
        as she sings her son's doom:
he was killed and she by her
own hand unmothered.

But as it happened, Niobe was also unmothered, losing five of her daughters and five of her sons because she had dared to question popular devotion to Leto, mother of only two children, never mind that the two were none other than Apollo and Artemis. Such an insult could not be brooked. Apollo shot the sons as they hunted in the mountains, and Artemis sent arrows through the daughters as they sat with their distaffs spinning wool in the palace. The two that survived were spared because they had delivered placatory prayers just in time.

But the male house wren opens his bill, puffs out his brown-striped breast, and sings not in lamentation but rather to proclaim the bounds of his fiefdom and to declare his success at enticing a mate into one of the cavities that he has selected. The pair has no idea that I rejoice in their presence, nor would they care if they could know. Our relationship, however, is a commensal arrangement in which both of us benefit substantially. I enjoy the music, of course, but that is just the icing on the cake. It's my garden that forms the bond between us and, more precisely, the bugs

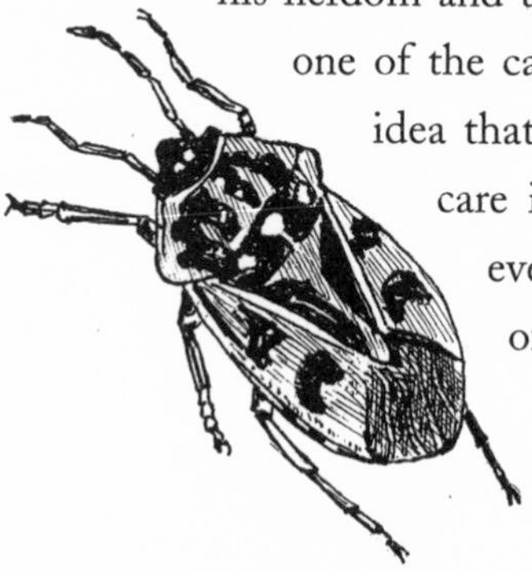

that my garden attracts. For, the little birds are ferocious insectivores, gobbling up cucumber beetles, two-lined spittlebugs, the harlequin bugs that infest the horseradish, and even the Japanese beetles that attack the strawberries and pole beans. Their cleansing efforts provide both of us with food—bugs for the birds, beans, cucumbers, and horseradish for me. And I think fondly of Bess, whose flowing signature brought the birdhouse, the house wrens, and their music to my yard.

Them and Us

{ *An Afterword* }

Groundhogs, mice, cottontail rabbits, cats, white-tailed deer, house wrens and house sparrows, chickens, turtles, slugs and snails, carpenter bees, harlequin bugs, bacteria, and even some plants, morning glories and the red maple—not one of my subjects is exotic. And here, in the end, I return to the book's beginning—to the complicated and not fully answerable question of cognition and consciousness in non-human creatures.

With prokaryotes, protoctistans, plants, and fungi, it can be said with certainty that they conduct biochemical communications internally—and externally, as well, when they deal with their compeers, competitors, and collaborators. Cognition and consciousness, however, cannot be detected. So, we sauté mushrooms and poison microbes with antibiotics. But with the animals, be they bugs, fish, or German shepherds, our views of their mental processes profoundly affect the ways in which we treat them. Bugs—especially mosquitoes and cockroaches—are to swat or squash, fish to catch and eat, German shepherds to serve as friends and bodyguards. Some philosophers and scientists wish to distinguish between perceptual and reflective consciousness, the former being active awareness of both body and the environment, and the latter, a simple consideration of one's internal weather. As we've seen in the case of the interior-decorator mice, such introspection may take the form of avoiding consumption of a product that had a bad taste or caused a tummy ache. The two modes of consciousness have also been described as simultaneous and ordinary. The simultaneous mode looks outward, and the barriers between self and the rest of the world dissolve. This, I believe, is a path often taken by birds, reptiles, and fish, which register all the signals that their surroundings deliver but respond on the instant if one signal should beckon or alarm. Ordinary perception looks inward: what do my surroundings do for me? What do I stand to gain or lose? It is a selective, self-referential way of responding to objects and circumstances. If it rains, I don't need to water the garden. If a groundhog appears amid my flowers and vegetables, I set the live-trap or hire the Critter Git'r. If I buy a lottery ticket, I have a chance to win a million dollars. But, as we've seen, cats, too, have a self-referential consciousness, and it's fair to say that not just our anthropoid kin but also the great gaggle of warm-blooded animals operates in the same way. And having met Wumpus the turtle, I am convinced that cold-blooded animals exhibit consciousness. She has learned her way around the house, and she asks to have her shell scratched. What's more, if one creature sees another member of its species flying or leaping or yawning, it is surely aware that a particular individual is performing those actions, and it is likely aware that it, too, can fly, leap, or yawn.

Bernard J. Baars, a noted neuroscientist, has said: "All animals engage in purposeful action . . . seeking food, mates, and the company of others. . . . Animals investigate novel and biologically significant stimuli as we do, ignore old and uninteresting events just as we do, and share our limited capacity for incoming information. We are not the only conscious beings on earth." I would add another element to his food, mates, and company trio: seeking to protect our offspring.

These statements hold moral and ethical challenges. If we find curiosity, playfulness, an ability to solve problems, and purpose in nonhuman animals, if we admit that they are conscious beings, then what should be the relationships between Them and Us? How should we treat them who have as much claim to life as we do? If we were Jains, we would be committed to non-violence and to the belief that all life-forms are equal—thermophile, moss, death angel, clam, ladybug, starling, bullfrog, orangutan, and every other living being in the world's grab bag. But most of us are not Jains, and we do not try to avoid killing other animals. Animal liberationists like the ethicist Peter Singer maintain that all animals have innate capacities for suffering and for enjoyment. From that premise, it follows that, because people can control the treatment that we deal out to them, we should behave like Jains and abstain altogether from cruelty and violence toward the other members of our kingdom. Away with factory farms and slaughterhouses! Close the labs that run experiments on chickens, mice, and monkeys! Become a vegetarian (despite dentition that marks us as omnivores)! We can certainly make farm animals more comfortable and work for clean labs with healthy stock. Yet, those are not panaceas, for human lives are ever in thrall to other lives. We seek food, mates, the safety of our young, and the company of others, nor are the others always of our kind. They may be dogs, cats, horses, parrots, or pot-bellied pigs.

Alfred, Lord Tennyson's lines from *In Memoriam A. H. H.* come to mind. The poem, published in 1850, sings his grief, a grief barely held in check, for his beloved friend Arthur Hallam, who died at the age of twenty-two. The poem is not just a drawn-out lamentation but also a lyric search for elusive peace and understanding. The lines that follow are not a brief for Darwinian natural selection and survival-of-the-fittest, for the poem appeared almost a decade before Darwin's *The Origin of Species*. They are, rather, a small part of Tennyson's attempt to come to terms with a punishing loss. I believe that they speak a plain truth about Them and Us. The "ravine" of which the poet speaks is an alternate spelling for "rapine."

Who trusted God was love indeed
  And love Creation's final law—
  Tho' Nature, red in tooth and claw
With ravine, shrieked against his creed.

Nothing has yet repealed Nature's dictates. We eat to live and kill to eat. So do all the world's other creatures. At rock bottom, we are Them and they are Us. There is no basic difference in our capacities for suffering and pleasure. But we have opportunities that they do not: the ability to reflect on what we do, the power to act humanely, the obligation not to waste a whit of what we take, and the grace to be thankful.

Notes

EPIGRAPH

v The real: Davis, 91.

THEM AND US: AN INTRODUCTION

xv One study: Ossinger.

xv "Cognitive potential": Gould, 178.

xvi At the cognitive level: Gould, 222.

xviii One bacterium: Paddock.

ONE: SOPHIE

5 "They celebrate": Herodotus, 153.

6 "So many cats were mummified": Hubbell, 88.

TWO: A BIRD OF CONSEQUENCE

19 Those breeds were said to be: Varro, 472.

20 he suggests: Varro, 472–74.

22 "the only way": Randolph, 18.

20 Put the fowls: Randolph, 18–19.

23 "chicken breeding represents": International Chicken Genome Sequencing Consortium, 712.

24–25 Now that the human genome sequence: Schmutz and Grimwood, 680.

25 "For nearly every aspect": International Chicken Genome Sequencing Consortium, 712.

THREE: WHITETAILS

31 According to the United States: United States Department of Agriculture Wildlife Services.

31 a report issued in 2006: Metropolitan Washington Council of Governments, 35.

32–33 the deer, crowded: Virgil, Book Three, 369–375

33 "bigger than bulls": Lombardo and Rayor, 14.

33–34 the third labor: Graves, 472–73.

39 "Food plots": Virginia Department of Game and Inland Fisheries, "Plantings to Attract Deer."

42 "Feeding can cause . . . other wildlife": Virginia Department of Game and Inland Fisheries, "Frequently Asked Questions."

44 "Reflectors deflect": DeNicola and others, 22.

45 "STOP BOWBARISM": DeNicola and others, 12.

45 "win-win": Thurow.

FOUR: THE WOODCUTTERS

48 "his floures": Gerard, 1215.

51 the magnificent Bee: Fabre, 8.

51 "the sky's celestial gift": Virgil, Book Four, 1.

52 . . . some say: Virgil, Book Four, 219–227.

52 "mysteries of the palace of honey": Maeterlinck, 5.

52–53 There is the distressful: Maeterlinck, 6.

55 "The Xylocopae": Maeterlinck, 106.

55 *Status*: Berenbaum.

55 Studies show: Berenbaum, 182.

55 It delivers the opinion: Berenbaum, 10.

57 "If you do not tease": Fabre, 83.

58 When the hour: Fabre, 38.

FIVE: THE INSIDE STORY

65 "The *E. coli* toxin": Liebman, 4.

SIX: RASCAL

70 "A morning-glory": Whitman, "Song of Myself: 24"

76 One garden historian: Pennington, 110.

77–78 "The Potato roots"; "Howsoever": Gerard, 926.

79–80 Somewhat of a rascal: Keeler, quoted by Adams, 144.

80 I send herewith: Betts, 155.

80 In March: Betts, 161.

SEVEN: A BALE OF CHELONIANS

87 "The tortuge": *Oxford English Dictionary*, under the entry for "tortoise."

87–88 "Summer-Country": Lawson, 70.

89 The old Sussex tortoise: White, 217–18.

89 "More than two-thirds": White, 218.

89 "abject reptile"; "poor embarrassed reptile": White, 219.

89 A modern riff: Klinkenborg, 7.

94 "all animals commonly known"; "viable turtle eggs"; "in a humane manner": Coy.

95 "An unintended consequence": *The Vivarium Magazine*, quoted in the Mid-Atlantic Turtle and Tortoise Society Newsletter, 5.

EIGHT: SPARROW CRIMES

107 "a shy, secretive": Farrand, 1, 226.

109 "the roof": Nelson, 24.

112 an indoor house sparrow: Tarte, 151–52.

NINE: MELANODON'S CHILDREN

122 "sets up housekeeping": Virgil, Book 1, 181–82.

126 "The mouse genome": Dalke.

126 "Though many human and mouse genes": Dalke.

129–30 "Since we owe them": Lawlor.

131 "Would it not be difficult": Collins, 90.

TEN: RED IN EVERY SEASON

139 "the Maple's poor relations": Peattie, 473.

139 "this bastard maple": Coues, quoted by Peattie, 474.

139–40 "may be rank'd"; The *Indians*: Lawson, 111.

140 "The Maple": Lawson, 105.

142 Out of its wood: Kalm, 88.

ELEVEN: BIG BEAST

145 "marvelous mammal": http://www.hoghaven.com/.

146 "For as the sun shines": http://www.stormfax.com/ghogday.htm.

TWELVE: THE MOLLUSKS IN THE GARDEN

158 "In 1857": Binney, quoted by the Western Society of Malacologists, 23.

159 Researchers report: Brosi.

163–64 "When the pursuer"; "The circle now grows": Adams.

165 Our cellar mollusks: Binney, 167.

THIRTEEN: MR. McGREGOR'S NEMESIS

177 "The cultivation of Rabbits"; "A Rabbit's borough": Simmons, 9.

177 "Rabbit can be prepared": Cameron, 97.

FOURTEEN: THE HARLEQUINS

182 The microscope: Davis, 83–84.

183 "a black mark"; "trademark": Davis, 82.

184 Look carefully: Davis, 86.

184 Fabre cites: Davis, 89–90.

186 "scabs, scares": Pliny quoted by Gerard, 242.

186 The leaves boiled: Gerard, 242.

FIFTEEN: A GIFT FROM A GHOST

191–92 will think that he listens: Aeschylus, 25.

THEM AND US: AN AFTERWORD

196 "All animals engage": Baars, quoted by Griffin, 284–85.

197 Who trusted God: Tennyson, *In Memoriam A. H. H.*, Canto LVI.

Bibliography

Adams, Denise Wiles. *Restoring American Gardens: An Encyclopedia of Heirloom Ornamental Plants, 1640–1940*. Portland, Oregon: Timber Press, 2004.

Adams, Lionel E. "Observations on the Pairing of *Limax maximus*." [Internet] Available from: http://members.tripod.com/arnobrosi/lea.html.

Aeschylus. *Suppliants*. Translated by Janet Lembke. New York and London: Oxford University Press, 1975.

Berenbaum, May R. *Ninety-Nine Gnats, Nits, and Nibblers*. Urbana and Chicago: University of Illinois Press, 1990.

Berenbaum, May R., and others. *Status of Pollinators in North America*. Washington, D.C.: National Academies Press, 2006. [Internet] Available from: http://www.nap.edu.

Betts, Edwin Morris. *Thomas Jefferson's Garden Book, 1766–1824*. Philadelphia: American Philosophical Society, 1992.

Binney, W. G. "The Mollusks in Our Cellars," *The American Naturalist*, Vol. 4, No. 3 (May 1870), pp. 166–71.

Brosi, Arno. "The Trail of the Snail." [Internet] Available from: http://members.tripod.com/arnobrosi/snail.html.

Cameron, Angus, and Judith Jones. *The L. L. Bean Game & Fish Cookbook.* New York: Random House, 1983.

Collins, Billy. *Sailing Alone Around the Room: New and Selected Poems.* New York: Random House, 2001.

Coy, Thomas. "Editorial: The 4-inch Law." Austin's Turtle Page. [Internet] Available from: http://www.austinsturtlepage.com/Articles/4inchlaw.htm.

Dalke, Kate. "Mouse in the House: Scientists Compare Mouse and Human Genomes." Genome News Network, December 4, 2002. [Internet] Available from http://www.genomenewsnetwork.org/articles/12_02/mouse.shtml.

Davis, Linda, ed. *The Passionate Observer: Writings from the World of Nature by Jean-Henri Fabre.* San Francisco: Chronicle Books, 1998.

DeNicola, Anthony J., Kurt VerCauteren, Paul D. Curtis, and Scott E. Hygstrom. *Managing White-Tailed Deer in Suburban Environments.* Cornell Cooperative Extension, 2000. [Internet] Available from: http://www.dgif.virginia.gov/wildlife/deer/suburban.pdf.

Fabre, J.-Henri. *The Mason-Bees.* Translated by Alexander Teixeira de Mattos. Garden City, New York: Garden City Publishing Company, 1935.

Farrand, John, Jr., ed. *The Audubon Society Master Guide to Birding, 1: Loons to Sandpipers; 3: Old World Warblers to Sparrows.* New York: Alfred A. Knopf, 1983.

Feduccia, Alan, ed. *Catesby's Birds of Colonial America.* Chapel Hill and London: University of North Carolina Press, 1985.

Gerard, John. *The Herball, or Generall Historie of Plantes: The Complete 1633 Edition as Revised and Enlarged by Thomas Johnson.* New York: Dover Publications, Inc., 1975.

Gould, James R., and Carol Grant Gould. *Animal Architects: Building and the Evolution of Intelligence.* New York: Basic Books, 2007.

Graves, Robert. *The Greek Myths: Complete Edition.* London: Penguin Books, 1960.

Griffin, Donald R. *Animal Minds: Beyond Cognition to Consciousness*. Chicago and London: University of Chicago Press, 2001.

Haughton, Claire Shaver. *Green Immigrants: The Plants That Transformed America*. New York and London: Harcourt Brace Jovanovich, 1978.

Heiser, Charles B. *Weeds in My Garden: Observations on Some Misunderstood Plants*. Portland, Oregon: Timber Press, 2003.

Herodotus. *The Histories*. Translated by Aubrey de Sélincourt. Harmondsworth, Middlesex, England: Penguin Books, Ltd., 1954.

Hubbell, Sue. *Shrinking the Cat: Genetic Engineering Before We Knew About Genes*. Boston: Houghton Mifflin Company, 2001.

International Chicken Genome Sequencing Consortium. "Sequence and Comparative Analysis of the Chicken Genome Provide Unique Perspectives on Vertebrate Evolution," *Nature*, Vol. 432 (December 9, 2004), pp. 695–716.

Jaeger, Edmund C. *A Source-Book of Biological Names and Terms*, 3rd ed. Springfield, Illinois: Charles C. Thomas, 1955.

Kalm, Peter. *Travels in North America: The English Version of 1770 Edited by Adolph B. Benson*. New York: Dover Publications, 1987.

Klinkenborg, Verlyn. *Timothy; Or, Notes of an Abject Reptile*. New York: Alfred A. Knopf, 2006.

Lawlor, Monica. "A Home for a Mouse." [Internet] Available from http://pyseta.org/hia/vol8/lawlor.html.

Lawson, John. *A New Voyage to Carolina*. Edited by Hugh Talmage Lefler. Chapel Hill: University of North Carolina Press, 1967.

Leighton, Ann. *American Gardens in the Eighteenth Century*. Boston: Houghton Mifflin, 1976.

Liebman, Bonnie. "Fear of Fresh: An Interview with Robert Tauxe," *Nutrition Action*, Vol. 33, No. 10 (December 2006), pp. 3–6.

Lombardo, Stanley, and Diane Rayor. *Callimachus: Hymns, Epigrams, Select Fragments*. Baltimore and London: Johns Hopkins University Press, 1988.

Maeterlinck, Maurice. *The Life of the Bee*. [Internet] Available from: http://www.gutenberg.org/etext/4511.

Metropolitan Washington Council of Governments. "Deer-Vehicle Collision Report September 2006." [Internet]Available from: http://mwcog.org/uploads/pub-documents/y1lywa20061030150157.pdf.

Mid-Atlantic Turtle and Tortoise Society Newsletter, March 2005.

National Human Genome Research Institute. "Background on Mouse as a Model Organism." [Internet] Available from: http://www.genome.gov/10005834.

Nelson, Dylan, and Kent Nelson, eds. *Birds in the Hand: Fiction & Poetry About Birds*. New York: North Point Press, 2004.

Ossinger, Joanna L. "Experiments Suggest Birds May Be Capable of Planning Ahead," *The Wall Street Journal*, April 13, 2007, p. B1.

Paddock, Catharine. "Soil Bacteria Work in Similar Way to Antidepressants," *Medical News Today*, April 2, 2007. [Internet] Available from: http://www.medicalnewstoday.com/healthnews.php?newsid=66840.

Peattie, Donald Culross. *A Natural History of Trees of Eastern and Central North America*. Boston: Houghton Mifflin Company, 1991.

Pennington, Susan J. *Feast Your Eyes: The Unexpected Beauty of Vegetable Gardens*. Berkeley, Los Angeles, London: University of California Press, 2002.

Pliny. *Natural History*, Vol. III. Translated by H. Rackham. Cambridge, Massachusetts: Harvard University Press, 1967.

Randolph, Mary. *The Virginia Housewife or, Methodical Cook*. New York: Dover Publications, Inc., 1993.

Rose, Kenneth D. *The Beginning of the Age of Mammals*. Baltimore: Johns Hopkins University Press, 2006.

Schmutz, Jeremy, and Jane Grimwood. "Genomics: Fowl Sequence," *Nature*, Vol. 432 (December 9, 2004), pp. 679–80.

Simmons, Amelia. *The First American Cookbook: A Facsimile of American Cookery, 1796*. New York: Dover Publications, Inc., 1984.

Singer, Peter. *Animal Liberation*. New York: HarperCollins Publishers, Inc., 2002.

Spencer, Colin. *The Vegetable Book: A Detailed Guide to Identifying, Preparing, and Cooking over 100 Vegetables*. New York: Rizzoli, 1995.

Tarte, Bob. *Fowl Weather*. Chapel Hill: Algonquin Books, 2007.

Thurow, Roger. "To Feed the Needy, Iowa Looks Hungrily at Its Glut of Deer," *The Wall Street Journal*, November 20, 2006.

U.S. Food and Drug Administration. *The "Bad Bug Book": Foodborne Pathogenic Microorganisms and Natural Toxins Handbook*. [Internet] Available from: http://vm.cfsan.fda.gov/~mow/intro.html.

United States Department of Agriculture Wildlife Services. "Living with Wildlife: Deer." [Internet] Available from: www.aphis.usda.gov/ws/nwrc/is/living/deer.pdf.

Varro. *On Agriculture*. Translated by William Davis Hooper, with a revision by Harrison Boyd Ash. Cambridge, Massachusetts: Harvard University Press, 1935.

Virgil. *Georgics*. Translated by Janet Lembke. New Haven and London: Yale University Press, 2005.

Virginia Department of Game and Inland Fisheries. "Plantings to Attract Deer." [Internet] Available from: www.dgif.virginia.gov/wildlife/deer/plantings.asp.

——. "Frequently Asked Questions." [Internet] Available from: www.dgif.virginia.gov/wildlife/deer/faq.asp.

von Muggenthaler, Elizabeth. "The Felid Purr: A Bio-Mechanical Healing Mechanism." [Internet] Available from http://www.animalvoice.com/catpurrP.htm.

Weaver, William Woys. *Heirloom Vegetable Gardening*. New York: Henry Holt and Company, 1997.

Western Society of Malacologists. *Field Guide to the Slug*. Seattle, Washington: Sasquatch Books, 1994.

White, Gilbert. *The Natural History of Selborne*. Oxford and New York: Oxford University Press, 1993.

About the Author

JANET LEMBKE is the author of many acclaimed books on natural history, gardening, and cooking, as well as numerous translations from Greek and Latin. Her essays have appeared in numerous magazines and journals, including *Audubon, Sierra*, and *The Southern Review.*